Jihen Elkhaldi
Lotfi Bouslimi
Mohamed Najeh Lakhoua

Modeling and Intelligent Energy Management

Jihen Elkhaldi
Lotfi Bouslimi
Mohamed Najeh Lakhoua

Modeling and Intelligent Energy Management

ScienciaScripts

Imprint

Any brand names and product names mentioned in this book are subject to trademark, brand or patent protection and are trademarks or registered trademarks of their respective holders. The use of brand names, product names, common names, trade names, product descriptions etc. even without a particular marking in this work is in no way to be construed to mean that such names may be regarded as unrestricted in respect of trademark and brand protection legislation and could thus be used by anyone.

Cover image: www.ingimage.com

This book is a translation from the original published under ISBN 978-620-6-71468-2.

Publisher:
Sciencia Scripts
is a trademark of
Dodo Books Indian Ocean Ltd. and OmniScriptum S.R.L publishing group

120 High Road, East Finchley, London, N2 9ED, United Kingdom
Str. Armeneasca 28/1, office 1, Chisinau MD-2012, Republic of Moldova, Europe
Printed at: see last page
ISBN: 978-620-8-11417-6

Modeling and Intelligent Energy Management

Jihen Elkhaldi

Lotfi Bouslimi

Mohamed Najeh Lakhoua

June 202 4

PREFACE

The rapid industrialization of recent decades and the proliferation of household electrical appliances (heating, air-conditioning, washing, medical, IT, etc.) have led to immense global demand for electrical energy. Today, more than 2 billion human beings have no access to electricity, due to fragile economies, heavy and costly infrastructures, difficult-to-access areas and scattered settlements [1].

What's more, global greenhouse gas emissions have risen by 30% since 1990, and energy consumption accounts for around two-thirds of global CO_2 emissions [2].

As a result, most states have adopted resolutions aimed at tackling the problems of depleting fossil fuel reserves in the future ("Europe 20-20-20" strategy, in 2009, the Union set itself the target of increasing the share of renewable energies in its energy consumption to 20% by 2020), reducing greenhouse gas emissions and polluting gases (Kyoto Protocol : this protocol aimed to reduce, through legally binding commitments, emissions of six greenhouse gases: carbon dioxide, methane, nitrous oxide and three chlorofluorocarbons substitutes, by at least 5% compared to 1990 levels), and to balance the increased demand for energy potential.

To meet these challenges, the use of environmentally-friendly energy sources has become an essential alternative for sustainable development [3].

It's true that renewable energy (RE) generation has always been present in distribution networks, but its presence has never been on a scale to have an impact on network operation. Today, however, the situation is beginning to change, with decentralized power generation increasing year on year [4].

However, their integration into conventional networks has led to other problems:

Renewable energy technologies are weather-dependent and suffer from a major drawback: their intermittency. To overcome this limitation, the use of hybrid systems combining multiple sources, such as renewable energy systems, the national distribution grid (historical power grid), conventional energy sources and storage systems, is generally considered an efficient and reliable solution for the future [5].

Most of today's power grids need to evolve to support the transition to an energy system based on renewable energies [6]. Smart Grids (SGs) are essential for this transition to an increasingly renewable energy mix, and for controlling rising energy demand.

An intelligent network is made up of a large number of interacting entities. It adapts to external or internal pressures to maintain its functionality, presenting multi-scale complexity in both space and time. Adequate modeling is essential to anticipate, analyze, discuss and predict, if possible, the system's behavior in order to determine the ideal model.

The challenges of Smart Grids are linked, on the one hand, to their complexity and, on the other, to the heterogeneity of the players involved, with divergent interests [7]. This makes the management of energy flows extremely complex. It is therefore necessary to develop algorithms adapted to these problems, in order to optimize energy management, guarantee smooth system operation and ensure continuous quality for consumers.

This book begins with a brief review of the impacts of climate change, as well as global energy issues and context. It then presents a general state of the art on hybrid systems, including their compositions, architectures, objectives, and some results from the literature review.

In the second chapter, the concept of microgrids is addressed, with a discussion of the various architectures and their main objectives. A detailed literature review on Smart Grids is also presented, covering their architectures, various domains, objectives and concluding with a list of Smart Grid research.

The third chapter offers an in-depth exploration of the different modeling methods used in the field, highlighting their advantages, disadvantages and specific areas of application. A detailed classification is also presented, enabling readers to better understand the diversity and scope of approaches available for modeling complex energy systems.

Finally, the last chapter is dedicated to an in-depth analysis of optimal energy management techniques. This chapter examines the methods and algorithms used to maximize energy efficiency, minimize costs and optimize the performance of energy systems.

Table of contents

Chapter 1:
Towards an Optimal Energy Mix: The Key Role of Hybrid Energy Systems

<u>Chapter 1</u>: Towards an Optimal Energy Mix: The Key Role of Hybrid Energy Systems

1. Introduction

An adjustment of electrical energy is necessary to limit CO_2 emissions, preserve the greenhouse, limit pollution, combat climate change and reinforce energy security.

The use of renewable natural energies, considered inexhaustible and clean, offers the best solution to all these environmental problems.

Most of these green energies are subject to the whims of nature. In this respect, the use of hybrid systems combining multiple sources is considered by all to be the solution of the future.

The first chapter begins with a brief overview of the impacts of climate change and the global energy context. This gives me a clearer picture of the challenges we face and the need for innovative solutions.

I then go on to give a general overview of hybrid energy systems. I examine in detail their composition, architectures and main objectives. This in-depth analysis highlights the key advantages of these hybrid systems, which judiciously combine different renewable and conventional energy sources.

This introduction, both contextual and technical, lays the foundations for understanding the central role of hybrid systems in the coming energy transition. It sets the scene for the following chapters, which will delve into specific aspects of these promising technologies.

2. Energy contexts and renewable energies

The global energy situation is constantly evolving, marked by environmental, economic and geopolitical challenges. Global energy consumption continues to grow as the world's population increases and emerging economies expand. However, this growth is also accompanied by a growing need to take account of environmental impacts, such as climate change, air pollution and the growing scarcity of natural resources. Energy security has become a major issue for many countries, as they seek to diversify their energy supplies to reduce their dependence on a single energy source. All these factors have contributed to a growing awareness of the importance of transitioning to cleaner, more sustainable sources of energy, such as solar, wind, hydroelectric and geothermal power. These energies, known as renewable energies (RE), are expanding rapidly and offer potential solutions to these environmental problems, but they still represent a small share of total energy production, putting further pressure on the world's energy resources and raising questions about the long-term sustainability of energy supplies.

Overall, the global energy context is complex and constantly evolving, with challenges to be met to ensure sustainable energy production and consumption for current and future generations, while preserving the environment.

2.1. Global energy context

Population growth, the race for growth and the development of a globalized economy have been accompanied by an explosion in energy consumption since the 1960s, and more specifically in electrical energy [4].

According to the IEA (International Energy Agency), global electricity demand rose by 4% in 2018 to over 23,000 TWh, contributing to a 20% growth in total final energy consumption [9].

The figure below shows that fossil fuels such as coal, natural gas and oil are the world's main sources of electricity generation.

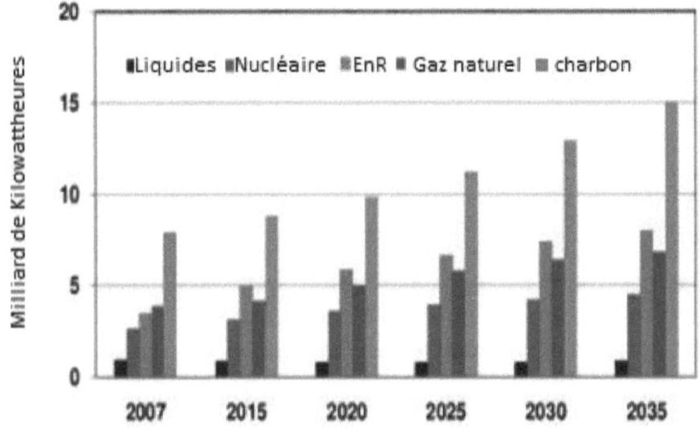

Figure 1 World net electricity production, 2007-2035 [10]

However, the consumption of fossil fuels is leading to a potential energy crisis in the future, as they are running out, and demand will soon outstrip production capacity. This makes the future dangerously fragile with this limited type of energy. What's more, fossil fuels generate greenhouse gas (GHG) emissions, which contribute to global warming.

From Figure 2, we can deduce that within 20 years, the global energy-related CO_2 emissions budget needed to keep warming below 2°C will be exhausted.

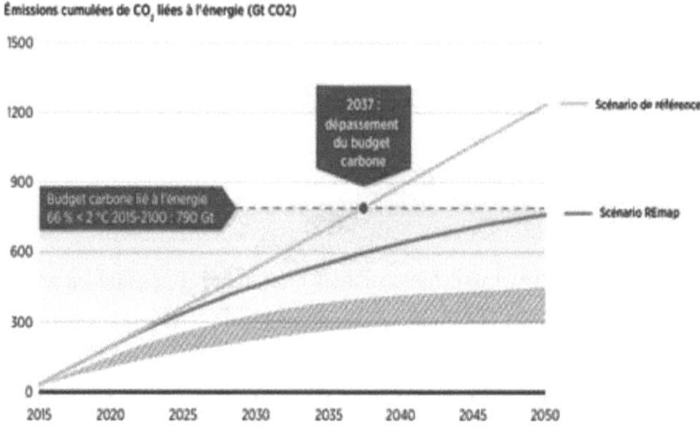

Figure 2Emissions and emissions gap, 2015-2050 [11]

A special report by the IPCC (Intergovernmental Panel on Climate Change) showed the impacts of a 1.5°C warming on natural and human systems, and analyzed the threat of climate change [12].

To address these issues, many countries and regions have taken energetic initiatives to increase their energy efficiency and renewable energy capacity. There has been a sharp increase in international agreements and national energy action plans.

At the United Nations Conference on Environment and Development (UNCED) in Rio de Janeiro in 1991, 150 countries made commitments under Agenda 21, underlining the importance of energy in sustainable development.

1997: The Kyoto Protocol is an international agreement on climate change adopted at the Conference of the Parties to the United Nations Framework Convention on Climate Change (UNFCCC). The main aim of the Kyoto Protocol was to reduce greenhouse gas (GHG) emissions from industrialized countries, particularly carbon dioxide (CO_2), in order to limit global warming and its harmful effects. The signatory countries undertook to collectively reduce their GHG emissions by at least 5% compared with 1990 levels.

1998: the United Nations Development Programme (UNDP), the United Nations Department of Economic and Social Affairs (UNDESA) and the World Energy Council (WEC) initiated the World Energy Report.

In 2001: 9ième session of the United Nations Commission on Sustainable Development (CSD-9), discussed the links between atmosphere and energy, as well as those between energy and transport.

In 2009: the European directive on renewable energies aims to increase the share of renewable energies in the European Union's energy consumption to 20% by 2020.

In 2015: the Kyoto Protocol was replaced by the Paris Agreement, which aims to limit global warming to less than 2°C above pre-industrial levels.

In 2015: the United Nations adopted the Sustainable Development Goals (SDGs), which aim to eradicate poverty, protect the planet and promote prosperity for all. The SDGs include specific targets linked to the fight against climate change and the use of renewable energies.

In 2018: the European Renewable Energy Directive was adopted, setting a binding target of 32% renewable energy in the European Union's gross final energy consumption by 2030. The directive also requires member states to promote the use of renewable energy sources, such as solar, wind and hydro power, as well as to foster energy efficiency.

Driven by these demands and motivations, renewable energy sources (RES) are being promoted to make energy production more reliable, cost-effective and environmentally friendly. The growth of renewable energy sources such as hydroelectricity, wind power, solar energy, geothermal energy, bioenergy, etc. has accelerated over the last decade.

2.2. Renewable energies

2.2.1. RE contexts

Renewable energies are energy sources produced from natural resources that are constantly renewed, such as sunlight, wind, water, geothermal energy and biomass. Unlike fossil fuels such as coal and oil, which are finite resources, renewable energies can be used sustainably without depleting natural resources. They are seen as more sustainable and environmentally-friendly alternatives to fossil fuels, which are finite and contribute to global warming.

Worldwide, renewable energy comes from six distinct sources. Hydropower is the main source, with a contribution of 82.9%. Biomass is the second source, with 6.3%. Next come wind power (8.3%), geothermal power (1.6%), solar power, which includes photovoltaic and thermal power plants (0.2%), and marine energy (0.01%) [13].

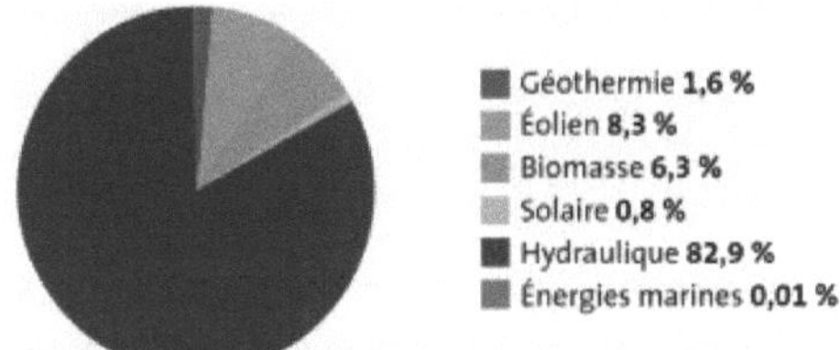

Figure 3Structure of renewable electricity generation [14].

According to the International Renewable Energy Agency (IRENA), over the period 2009-2018, RE capacity doubled, with 1221 gigawatts (GW) of RE added to the global power system [12].

Figure 4, shows the growth in renewable generation capacity. In 2018, solar continues to dominate as in 2017, with a capacity increase of 94 GW (+24%). It is followed by wind power with an increase of 49 GW (+10%) and hydro capacity up by 21 GW (+2%) [15].

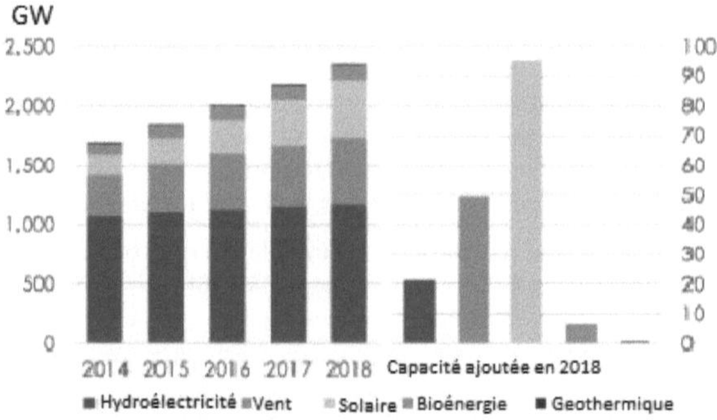

Figure 4Global capacity growth of renewable energy sources [12].

Figure 5 shows the share of renewable energies in the CTEF (%) and its projection over the next two decades. The total share of renewable energies must reach two-thirds by 2050 if climate targets are to be met.

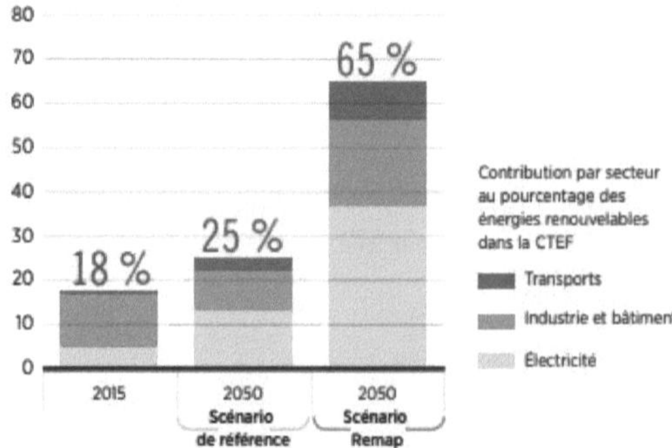

Figure 5Share of renewable energies in the CTEF (%) [11]

2.2.2. RE challenges

Renewable energies offer many advantages, particularly in terms of environmental sustainability and reducing greenhouse gas emissions. However, they also present a number of challenges:

- Intermittency: Renewable energy sources such as solar and wind are characterized by their intermittent nature, meaning that they do not provide constant electricity production. This underlines the importance of energy storage solutions to ensure a reliable power supply.

- High costs: Although the costs of renewable energy technologies have fallen considerably in recent years, they are still often higher than those of fossil fuels.

- Environmental impact: Renewable energy technologies have a different environmental impact than fossil fuels. Although they emit far fewer greenhouse gases, some renewable energies, such as large-scale solar power generation, can have environmental impacts on local ecosystems and natural habitats.

- Land use: Renewable energy projects often require major onshore or offshore installations, which can have an impact on land and sea use.

- Dependence on weather conditions: RE sources are often dependent on weather conditions, which can limit their use in certain geographical areas.

- Infrastructure constraints: Electricity grids need to be designed and updated to integrate intermittent renewable energies, which can be costly and requires long-term planning.

To further encourage the integration of renewable energies and meet the challenge of variable and unguaranteed power, an approach that couples different sources of supply and forms a hybrid system offers a promising solution.

3. Hybrid systems electrical energy

The integration of renewable energies into power grids plays an essential role in achieving greenhouse gas emission reduction targets, improving energy security and reducing electricity costs. However, it is equally important to guarantee the stability and reliability of the power grid.

These renewable energy sources have major advantages, including sustainability, low CO_2 emissions and economic benefits. However, most of these energy sources behave intermittently as a direct consequence of weather conditions. However, the influence of their intermittent nature can be mitigated by coupling two or more renewable sources, or with other conventional sources, for grid injection or isolated load supply [16].

3.1. Definitions

A hybrid system (HS) is a system that combines two or more different technologies, energy sources or operating methods to achieve a common goal. In general, HS systems seek to combine the advantages of each technology or energy source to create a more efficient, effective and profitable system.

A hybrid energy system (HIS) is an electrical system comprising more than one energy source, at least one of which is renewable. The HES may include a storage device. From a more global point of view, the energy system of a given country can be considered as a HS. [17], [18], [19].

Figure 6 shows the general architecture of a multi-source hybrid energy system.

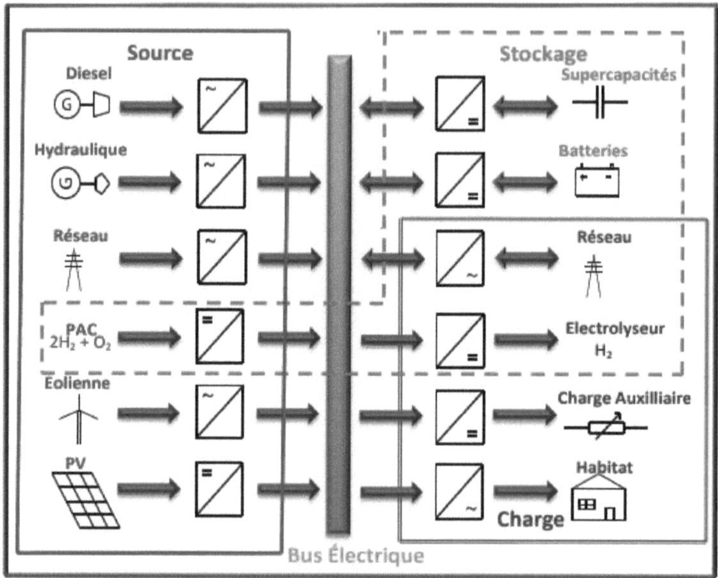

Figure 6General diagram of a multi-source hybrid energy system: loads and storage [20].

SHs offer many advantages and are becoming increasingly important in meeting current and future energy needs.

3.2.SHE components

The main components of an SHE depend on the application, but can include:

- Power sources: these provide the energy needed to operate the system. They include renewable energy sources (photovoltaic, wind and hydro generators, etc.) and conventional sources (diesel generators, batteries, etc.).

- Energy storage system: a storage device is an essential element in a multi-source system. Without it, the site's autonomy cannot be guaranteed. This device acts as an energy reservoir when the sources produce a surplus, and releases it when the sources don't produce enough. But beyond this energy reservoir aspect, a storage device is essential for the balance of power in a power grid [12]. There are many different technologies and types of storage, such as batteries, supercapacitors, fuel cells and so on.

- Energy conversion system: converts the energy produced by power sources into a form usable for the application. Frequently encountered converters are rectifiers (perform AC/DC conversion to charge batteries), inverters (convert DC to AC) and choppers (perform DC/DC conversion to adapt the voltage between two sources).

- Energy management system: manages the supply and distribution of energy in the SH. It can include load control devices, load controllers, power controllers, sensors, monitoring systems, etc.

- Energy distribution system: distributes the energy produced by energy sources and stored in energy storage systems to loads. Cables, switches, circuit breakers, transformers, electrical distribution networks, etc. are examples of components used in power distribution systems.

- Load: in a HS, load refers to the energy demand of the application using the system. This load makes electrical power useful, is variable and depends on the application. In a SHE, the energy management system monitors the load and selects the appropriate energy sources to meet the energy demand.

3.3.Study of different SHE structures

A number of SHE classifications are made, depending on the criterion chosen. Below, we present the most popular classifications.

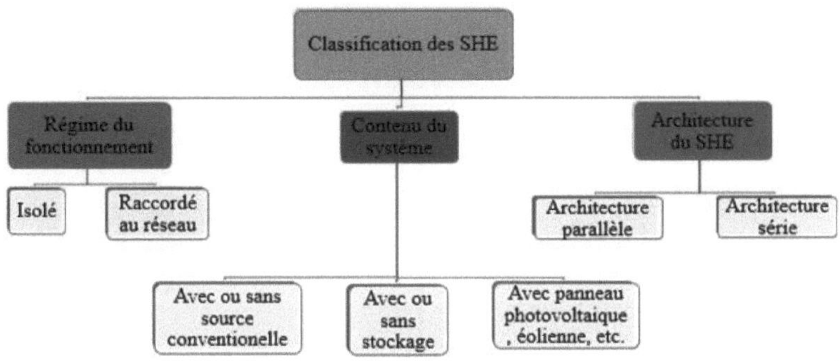

Figure 7SHE classification

In this section, we've chosen photovoltaic panels and wind turbines as SEnR and batteries for storage.

3.3.1. SHE connected to the network

A grid-connected SHE is an energy system that combines several energy sources to meet the energy needs of a community while being connected to the national power grid.

The majority of grid-connected SHs are without storage devices. Nevertheless, in some systems connected to a "weak" grid, storage units are introduced to reinforce the grid in case of failure [20].

Energy generated from renewable sources is used locally to meet the community's energy needs, while excess energy is fed into the national grid. In the event of an energy deficit, the community can draw electricity from the national grid.

The benefits of a grid-connected SHE include greater flexibility in managing energy production and distribution, reduced dependence on fossil fuels, lower fuel transportation costs, reduced greenhouse gas emissions, and improved energy security and resilience in the event of a natural disaster or disruption to the national power grid.

However, its implementation can require significant investment in the infrastructure and technology needed to effectively integrate different energy sources and manage power generation and distribution.

3.3.2. Stand-alone SHE

A stand-alone SHE operates totally isolated from the public grid. It is an energy system that combines different energy sources to meet the energy needs of a community or isolated geographical area.

The role of complementary sources is to assist the main sources by supplying the shortfall in renewable power generated. Batteries can also be used to store surplus renewable energy and release it when needed [16].

Isolated SHE is generally used in remote areas where connection to the national grid is impossible or too costly. It can also be used in areas where energy supply is intermittent or inadequate. The isolated SHE can be designed to meet the community's energy needs autonomously, depending on locally available energy sources.

Most remote sites, whether mining or community sites, have good wind and solar resources [21].

Figure 8 shows a typical example of an autonomous hybrid production unit that combines two energy sources, wind and solar, with an energy storage system.

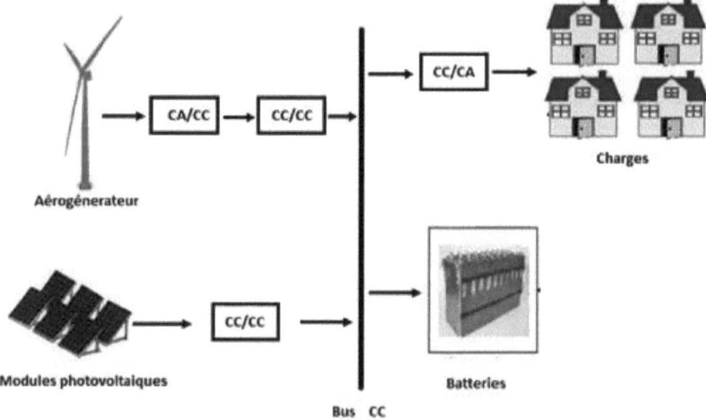

Figure 8Example of an SHE configuration dedicated to an isolated site

3.3.3. SHE with conventional source

In a conventional SH, a conventional energy source is combined with a renewable energy source.

The main advantage of a SHE with a conventional source is that it can provide reliable, stable power even when the RE source is unavailable or insufficient. This can be particularly important in regions where renewable energy sources are intermittent or unreliable due to variable weather conditions.

However, conventional energy sources have significant environmental impacts, including emissions of greenhouse gases, air pollutants and hazardous waste. To minimize these impacts, it is important to choose conventional energy sources that are as clean as possible, and to maximize the use of available renewable energy sources.

3.3.4. SHE without conventional source

These SH operate mainly in stand-alone mode at sites where diesel fuel supply or connection to the power grid is difficult or even impossible [22].

In addition, an SHE without a conventional source can also include energy storage batteries to store the energy produced during periods of surplus production and release it when demand is high.

This type of system offers many advantages, including reduced dependence on fossil fuels, lower long-term operating costs, lower greenhouse gas emissions and the creation of local jobs in the renewable energy sector.

However, it's important to note that setting up such a system requires careful planning and technical expertise. It's also important to take into account local environmental and geographical conditions to determine the most appropriate SEnR for each situation.

3.3.5. SHE with storage

The interconnection of the storage device with the energy sources can serve two possible purposes: either as a buffer when the system operates in parallel with the power grid, in which case the device smoothes out rapid variations in power, or as a longer-term, stand-alone storage device to supply the consumer when energy production decreases [22].

Energy storage is often provided by batteries, electrolyzers with hydrogen storage tanks, hot water storage tanks, and so on.

The main advantage of a hybrid energy system with storage is that it can provide a reliable and constant power supply, even when renewable energy sources are intermittent.

3.3.6. SHE with serial architecture

A SHE with serial architecture is a system that uses different SEnRs in series to meet energy needs. In this type of system, the energy sources are connected in series, so that the energy produced by one source feeds the next.

There are two configurations in this architecture: a DC bus configuration and an AC bus configuration.

3.3.6.1. DC bus configuration

In a series configuration, each source is individually connected to a DC bus using suitable power electronics converters, as previously illustrated in figure 8.

This means the system can supply AC loads or be connected to the grid via an inverter (DC/AC converter).

For low-power systems, the problem of frequency regulation is negligible. However, for medium- and high-power systems, DC bus voltage regulation remains particularly difficult [23].

The advantages of this structure can be summarized as follows:

-   Greater energy efficiency: this configuration enables more efficient conversion of renewable energy into usable energy. In fact, energy losses due to the conversion of energy into AC are reduced, enabling the RE to be used more efficiently.

-   A simpler control system,

- Less electromagnetic noise: DC bus topology can reduce the electromagnetic noise generated by the system. This can be particularly important in applications sensitive to electromagnetic noise, such as communication systems.

- Thanks to the inverter, the load can be supplied with the right voltage and frequency.

Although this configuration offers many advantages, there are also a few disadvantages to consider [16] :

- The efficiency of the whole system is relatively low, due to losses in the converters in particular.

- The inverter cannot work in parallel with the conventional source.

- A problem in the inverter can cause a complete power failure.

3.3.6.2. AC bus configuration

An AC bus architecture is a topology that uses an AC power distribution network to distribute energy generated by different energy sources, including renewable energy and conventional energy sources. In this architecture, energy generated by renewable energy sources is typically converted to AC by inverters before being distributed over the distribution network.

The batteries are connected to the AC bus via a bidirectional converter. DC loads can also be supplied from the batteries.

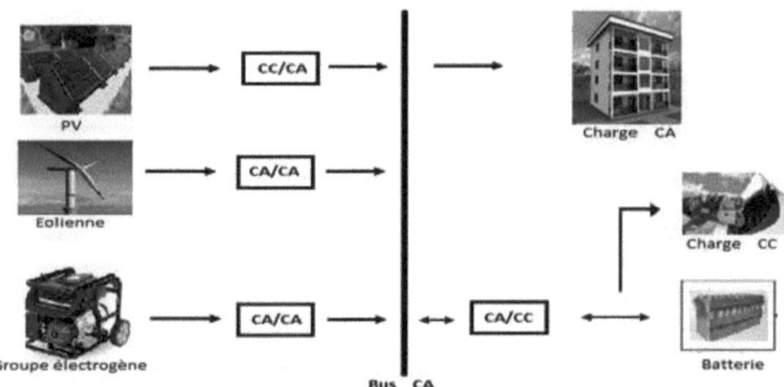

Figure 9Example of an AC bus SHE configuration

The selection of this configuration is appropriate as it enables direct AC voltage supply and guarantees continuous power supply to the load.

However, the disadvantage lies in the difficulty of imposing the frequency and amplitude of the bus voltage and adapting them to the load, as well as synchronizing all the modules in the network. In addition, a great deal of work needs to be done to shape the bus power. Consequently, this architecture is best suited to large networks with the necessary logistical resources [24].

3.3.7. SHE with parallel architecture

The parallel configuration uses two buses: a DC bus to which the DC sources (battery and PV panels) are connected, and an AC bus to which the AC sources (wind turbine, diesel generator, etc.) and the load are connected. The two buses are linked by a bidirectional converter [16].

The architecture of such a system is shown in Figure 10.

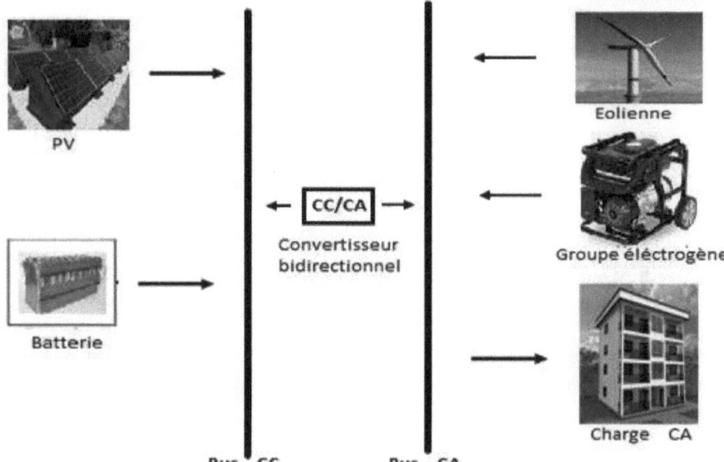

Figure 10Example of a parallel SHE configuration

This mixed structure combines the advantages of AC and DC bus structures. In this configuration, energy sources or electrical loads can be connected directly to the bus, with or without the power interface. As a result, the system will deliver greater energy efficiency at lower cost. However, the control and management of energy in this hybrid configuration is more complex than in conventional DC or AC configurations [25].

The bi-directional converter works either in rectifier mode, when the diesel generator fully covers the load demand and helps charge the battery, or in inverter mode, when the load is supplied by the panels and/or batteries. In this way, the load can be supplied by both buses simultaneously [26].

This system can offer a number of advantages, such as

- By using two different buses, the SH can improve system reliability and availability. If one bus fails, the other can continue to supply power to keep the system running.

- Reducing the number of converters.

- A problem occurring on a converter does not lead to the load's power supply being interrupted.

- The systems are able to satisfy both DC and AC consumption at the same time.

3.4.Research work on SH

In light of the above, it's easy to imagine the diversity of HS studies available.

Researchers V. Sark et al, (2004) [27] have studied an SH, consisting of photovoltaic panels and a diesel generator, in the Maldives, where solar potential is high. The aim of these systems, when operating in autonomous mode, is to supply a target load without interruption.

The addition of storage devices is imperative in this architecture as was indicated in the paper by researchers A. Adiyabat et al (2004) [28], they presented a techno-economic analysis of SHs for rural villages in Mongolia and compared the different types on the basis of net present cost and electricity cost for each load model case in villages in the Gobi region. As a result, they found that the existing diesel generation system ranked best even at high fuel costs in all load model cases.

Numerous researchers indicate that photovoltaic/diesel/battery SHs are reliable energy sources and represent an economically acceptable compromise between the high cost of stand-alone photovoltaic systems and the operation and maintenance and fuel costs of generators. This type of energy system is known to be one of the most cost-effective solutions for meeting the energy needs of remote regions [5].

Researchers R. Abhinav et al (2016) [29] studied SHs where the wind turbine presents a main source of energy, they separately addressed issues relevant to active and reactive power management to mitigate fluctuations. They gave an overview of the different control strategies and battery technologies applicable to each problem. These types of systems are commonly used on islands, where sea winds favor the use of wind power for electricity generation.

Researchers A. Mahesh et al. (2015) [30] discussed photovoltaic/wind SHEs with battery storage, they explained the exploitation of such hybrid energy systems for further improvements in terms of designing, analyzing and integrating these systems into the power grid.

The use of a photovoltaic/wind/diesel SH can be more reliable for meeting electricity demand in remote areas than photovoltaic/wind-only systems [31], [32].

Researchers A. Haghighat Mamaghani et al (2015) [33] analyzed the application of photovoltaic (PV) panels, wind turbines and diesel generators in autonomous hybrid energy production in a rural electrification system in three off-grid villages in Colombia with different climatic characteristics. The areas were selected according to Colombia's "Development Plan for Non-Conventional Energy Sources". First, different combinations of wind turbine, PV and diesel generator are modeled and optimized to determine the most energy-efficient and cost-effective configuration for each location. HOMER software was used to perform a techno-economic feasibility of the proposed SHs, taking into account net present cost, initial cost of capital and cost of energy as economic indicators.

Many researchers report that SH photovoltaic/wind/diesel/battery systems are reliable energy sources and represent an economically acceptable compromise between the high cost of stand-alone photovoltaic systems and the operation and maintenance and fuel costs of generators. This type of energy system is known to be one of the most cost-effective solutions for meeting the energy needs of remote regions [34].

4. Conclusions

Hybrid energy systems play a major role in the transition to a more sustainable and efficient energy future. By strategically integrating various sources of renewable and traditional energy, as well as storage technologies, these hybrid systems offer a number of significant advantages.

Thanks to their complementary nature, they can better manage the variability inherent in renewable energies. They also optimize the use of available energy resources, improving the reliability and security of supply of the power grid. And best of all, their environmental footprint is significantly reduced.

<u>Chapter 2:</u>

Microgrids and the Evolution of Smart Grids

<u>Chapter 2</u>: Microgrids and the evolution of Smart Grids

1. Introduction

In order to substitute renewable energies for traditional fuel-based energy applications, it is imperative that most of today's power grids evolve to support this transition. In this context, this chapter will focus on two key elements of this evolution: microgrids and smart grids.

Microgrids, as autonomous systems capable of operating independently or as a complement to the main grid, appear to offer greater flexibility and resilience. In fact, they enable more efficient integration of renewable energy sources.

At the same time, smart grids harness digital technologies to optimize management of the power system as a whole. Sensors, bidirectional communications and advanced control systems make it possible to better balance supply and demand, massively integrate variable renewable energies and actively involve consumers.

2. Smart microgrids

The architecture of smart grids continues to develop. And according to current research, these networks will comprise millions of parts and components: controls, computers, power lines, as well as new technologies and equipment. Consequently, the complexity and size of these networks will make the development of smart grids very costly and difficult, and it will take some time for all the technologies to be perfected, the equipment installed and the systems tested before they can be put into operation. And, in order to bring these networks into the practical realm of reality, the idea of creating interconnected smart micro-grids connected to public networks was put forward as an excellent solution. That's why we're looking to integrate them into buildings, or even neighborhoods, as they adapt to the development of renewable energies.

2.1. Definition

A microgrid is a discrete energy system composed of distributed energy sources (demand management, storage and generation) and loads capable of operating in parallel or independently of the main power grid. The main objective is to ensure local, reliable and affordable energy security for urban and rural communities [35].

R. H. Lasseter proposed the first microgrid architecture called Clean Energy Resources Teams (CERTS) [36], [37]. The CERTS microgrid generally assumes distributed generation units interfaced with converters based on RE and non-renewable sources.

As shown in Figure 11, renewable energy systems, energy storage systems and loads are connected to the main grid via power converters, with appropriate protection, communication, monitoring and control systems. It can operate in either grid-connected or islanded mode, subject to the operational characteristics of the main grid [38].

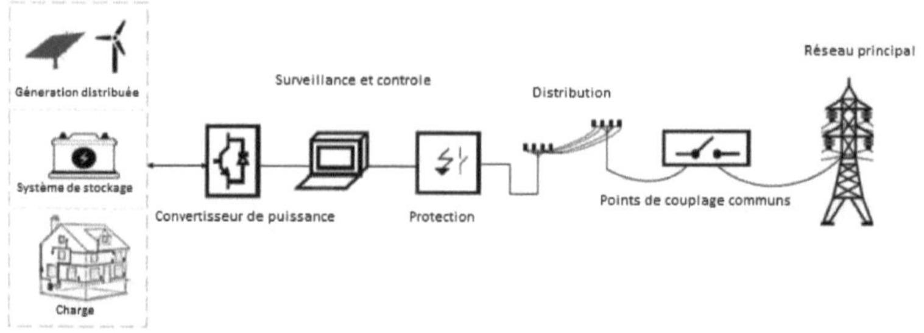

Figure 11A typical microgrid architecture [39].

Microgrids are small versions of the traditional power grid. Like current power grids, they support power generation, distribution and control (voltage regulation, switchgear). However, microgrids differ from traditional power grids by offering greater proximity between power generation and power consumption, resulting in increased efficiency and reduced transmission [40].

Another advantage is that microgrids can also be integrated with renewable energy systems, minimizing carbon footprint and greenhouse gas emissions.

Finally, microgrids provide dynamic control of energy sources, enabling autonomous and automatic self-healing operations. During normal or peak use, or at times of primary power supply grid failure, a microgrid can operate independently and is capable of feeding power back into the main grid [41].

2.2. Microgrid structure

The structure of an autonomous microgrid varies according to the constraints of the application. Nevertheless, four elements are often used [42]:

> Sources (renewable, batteries, diesel generator, etc.) ;

> Power converters (for system control and management) ;

> Interconnection filters (to ensure transmission of quality power) ;

> Loads (static or dynamic).

In addition, microgrids feature both AC and DC distribution lines. Each type has its own transmission and distribution levels. Common coupling points form the gateway between the main grid and the microgrid.

Microgrids frequently feature protection mechanisms to ensure that the system operates according to different principles and parameters.

Given the nature of the voltage and current, microgrids are sub-categorized into three types, namely AC bus, DC bus and mixed bus configurations. Each configuration has distinctive features, offering different advantages and disadvantages that need to be taken into consideration.

2.2.1. DC bus-coupled microgrid

This configuration is known as "Stand-Alone", and is a natural solution for combining most of the DC sources used in this type of application (photovoltaic panels, fuel cells, storage elements, etc.) [43], [44], [45].

Here are just a few of the advantages:

> ➢ In the DC system, fewer power converters are required, enabling size optimization and improved overall efficiency.
> ➢ Natural and easy to use.
> ➢ Extensive feedback on storage element lifetimes.

Disadvantages :

> ➢ The DC distribution protection system is not mature enough compared to the AC system.
> ➢ All the power is routed through the DC/AC converter, resulting in over-dimensioning of the inverter, which is responsible for supplying only the necessary power peaks.

2.2.2. AC bus-coupled microgrid

This configuration is based on the AC bus, and all variable-frequency sources (e.g. wind turbines) are connected to the bus via AC/AC power converters. Sources with a DC output (e.g. photovoltaic panels) are connected via DC/AC converters.

Although this structure is less common than the first, it has several advantages, the most important of which are [42] :

> ➢ The DC/AC converters associated with the source are sized according to the maximum power of the renewable source, rather than that of the load, which is less restrictive.
> ➢ Respond to load calls without going back to the DC bus, which is the case with DC hybridization. This would increase efficiency (reduced losses from the main inverter).
> ➢ Offer an additional degree of freedom for reagent control, an option that was not available with the previous configuration.

Its disadvantages include

> ➢ An AC bus microgrid requires a relatively complex controller to import and export energy while maintaining system stability and reliability.
> ➢ Poor impact on the lifespan of storage elements (batteries).

2.2.3. Mixed bus microgrid

The most feasible solution suggested by many researchers is to build a mixed AC/DC bus microgrid system for better integration of all elements with the main grid [46], [47]. This idea presents a new paradigm for defining distributed generation operation, in which AC and DC sources are connected to corresponding networks.

2.3. The objectives of a micro network

The objectives of a micro network may vary depending on the application and the environment in which it is used, but here are some common objectives:

> ➢ Providing reliable energy: microgrids are often used in rural or isolated areas where the traditional electricity grid may be unstable or non-existent. One of their main objectives is to provide a reliable source of energy for users.
>
> ➢ Reduce costs: they can often be cheaper to set up and manage than traditional electricity grids. By using renewable energy or energy storage technologies, they can also reduce long-term operating costs.

- ➢ Improve safety and resilience: they are designed to operate autonomously, even in the event of a power failure or natural disaster. This can improve network security and resilience, as well as power supply reliability for users.

- ➢ Reduce greenhouse gas emissions: by using renewable energies and optimizing energy management, a microgrid can reduce greenhouse gas emissions associated with energy production.

3. Smart grids

With population and economic growth, energy demand has increased significantly. This has led to increased pressure on existing power grids. At the same time, the negative environmental impact of fossil fuels, which are the world's main source of energy and which have been declining remarkably, has also increased.

In this process, we have had to meet growing demand by making optimum use of available energy resources, while integrating renewable energy sources that often operate intermittently and unpredictably.

Decentralized electricity generation can have several impacts on the power grid:

- ➢ Generation variability: Renewable energies are often intermittent, and this variability can cause voltage and frequency fluctuations on the power grid, which need to be monitored and managed to ensure grid reliability.

- ➢ Insufficient transmission capacity: power grids were designed to transport electricity from power plants to consumers. Distributed generation can reverse the flow of electricity through the grid, leading to saturation of existing transmission lines. Grid operators must therefore plan investments in transmission infrastructure to meet this growing demand.

- ➢ Increased transmission losses: this type of generation can lead to additional transmission losses, as electricity has to be transported over longer distances and through more conversion points. This can affect overall network efficiency and increase costs for consumers.

- ➢ Generation coordination: coordinating the entire power system is becoming more complex, as many small producers may be involved. Grid operators must therefore set up coordination and control mechanisms to maintain grid stability.

- ➢ Balancing supply and demand: distributed generation can also make balancing supply and demand more difficult, as production can vary considerably depending on weather conditions and other factors. Grid operators must therefore put in place regulation mechanisms to adjust production in line with demand in real time.

The implementation of decentralized production systems necessarily requires real-time flow control, adjustment between electricity supply and demand, and two-way communication since the consumer also becomes a producer, which has a major impact on the power grid.

The move towards smart grids is essential to enable an effective and sustainable energy transition, by making more efficient and cleaner use of electrical energy. This requires close collaboration between the various players in the energy system, including producers, distributors, consumers and regulators, to ensure a harmonious transformation of the energy system.

3.1. smart grid concept

Faced with the growth in consumption and the massive development of decentralized production, the electricity networks of the future are expected to have the following qualities [48], [49], [50], [51] :

- Flexibility: the network must meet the needs of all customers, taking future developments into account.

- Accessibility: the network must be able to accommodate all users (centralized and decentralized producers, consumers) and enable highly energy-efficient local production.

- Reliability: the network must ensure and improve the safety and quality of the power supply.

- Economy: the network must guarantee optimized costs through innovation, efficient energy management, free competition and equal regulation for all users.

The term Smart Grid (SG) has been presented as a combination of energy, communications, software and automated protection hardware [52].

The definition of a SG is the integration of an electrical network, a communications network, software and hardware to monitor, control and manage the production, distribution, storage and consumption of energy. Taking into account the need to support the increase of system efficiency through the integration of renewable energies and the improvement of security of supply, as well as the reduction of tariffs related to the distribution network and control energy. This is achieved by controlling the various generation facilities and by monitoring load connections according to needs, using IT technologies to adjust the flow of electricity between suppliers and consumers [53], [54].

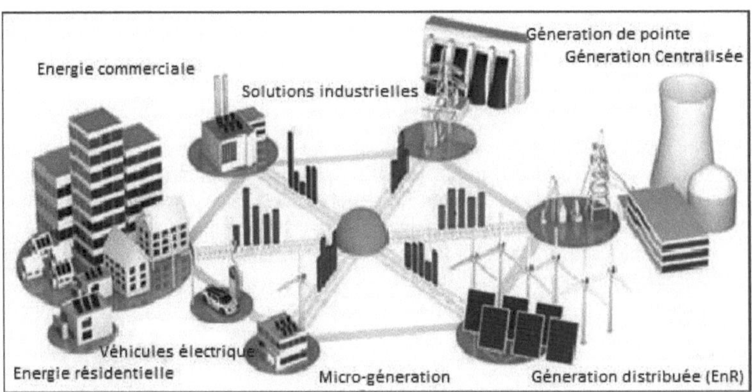

Figure 12Smart grid concept [54]

SGs, or smart grids, have been developed to improve the efficiency, reliability and sustainability of power grids using advanced information and communication technologies (ICT).

More specifically, SGs have been developed to solve the following challenges:

- ➢ Integration of renewable energies: SGs enable the efficient integration of renewable energies into the electricity grid. This reduces dependence on fossil fuels and promotes the transition to cleaner energy.

- ➢ Demand management: they enable efficient management of electricity demand in real time. Users can adjust their consumption according to electricity prices and demand signals to avoid power shortages and peak demand.

- ➢ Improved reliability: detection of outages and power quality problems will be faster, enabling utility providers to reduce downtime and improve the quality of power supplied to consumers.

- ➢ Cost reduction: EESs help to cut power grid operating costs by reducing energy losses, avoiding unnecessary investment in infrastructure and optimizing the use of available resources.

In short, EESs are used to meet the increasingly complex energy challenges of our time, by enabling more efficient and sustainable electricity management.

3.2. Smart grid architectures and domains

In this section, we present the different architectures and domains of a smart grid.

3.2.1. Smart grid architectures

The architecture of a GMS can be divided into three main levels:

- ➢ Infrastructure level: this level is responsible for managing the physical infrastructure of the power grid, including power plants, transformers, transmission lines, substations, distribution equipment and smart meters. Technologies used at this level include sensors, remote control and monitoring systems, switching and protection equipment, energy storage equipment and communication technologies.

- ➢ Information level: this level is responsible for managing the information generated by the power grid, including consumption data, RE generation data, energy storage data and power quality data. Technologies used at this level include information management systems, data management systems, Internet of Things (IoT) technologies, Big Data technologies and artificial intelligence technologies.

- ➢ Application level: this level is responsible for managing the interaction between the power grid and end users, including residential, commercial and industrial customers. Technologies used at this level include energy management systems, demand control applications, dynamic energy pricing applications, energy monitoring and control applications, RE generation applications and energy storage applications.

3.2.2. Smart grid domains

The SG is classified into seven domains which are presented in the following figure [55], [56]:

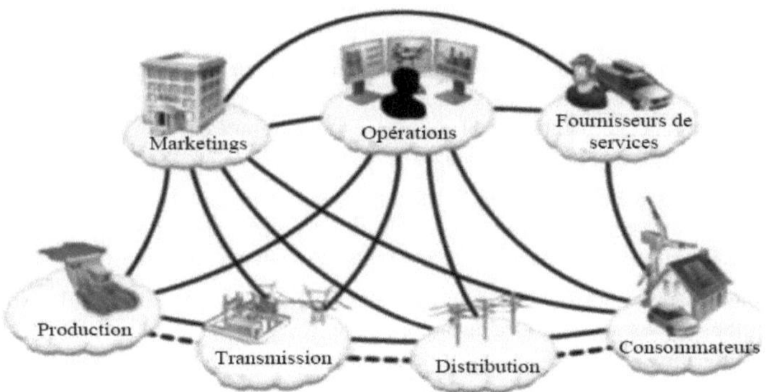

Figure 13The different domains of the smart grid [55], [57].

- Production area :

The generation area is a key element of a GXS, as it is responsible for producing the electricity needed to meet consumer demand for energy.

The transmission area is usually the boundary of the production area. The two areas are electrically connected.

Communications with the transmission domain are the most important, because without transmission, consumers cannot be served [58].

This field can include a variety of energy sources, such as :

- Conventional power plants.

- RE: such as solar, wind, hydraulic, geothermal or biomass energy.

- Energy storage sources.

- Microgrids: these local networks can be used to generate and distribute electricity in a specific community, using renewable energy and energy storage systems.

The main roles of the production area in a GMS include :

- Electricity supply: responsible for generating the electricity needed to meet consumers' energy demand, based on real-time requirements.

- Optimize production: optimize electricity production to minimize production costs, while guaranteeing a reliable power supply.

- Integrate renewable energies: it must integrate renewable energies, using energy storage technologies to respond to variations in production.

- Reduce greenhouse gas emissions: we must reduce greenhouse gas emissions by using cleaner technologies and increasing the share of renewable energies in electricity production.

In short, the generation area in a SG plays a key role in delivering reliable, sustainable and cost-effective energy to consumers. It must be able to produce energy according to demand, meet the coordination needs of the intelligent energy system, and promote renewable energies for a more sustainable future.

- Transmission area :

The transmission domain of a GXS is the section of the electrical network that ensures communication between the various components of the energy management system. This includes metering equipment, control systems, energy storage devices, renewable energy systems, smart metering devices and end-users.

Communication technologies used in the transmission domain of a GMS include wired and wireless communication networks such as fiber optic networks, Wi-Fi networks, cellular networks, powerline networks, and satellite communication networks.

The choice of technology will depend on a number of factors, such as the distance between the various network elements, population density, the presence of buildings or obstacles, the costs associated with setting up and operating the network, and so on.

In addition, the transmission domain of a SG also includes power quality control technologies, which ensure that electrical energy is supplied to consumers at the right voltage and frequency, while minimizing disturbances in the network.

- Distribution :

The main task of distribution networks is to deliver "medium and low voltage" energy from the transmission network to consumers over short distances [54], [59], [60].

Traditionally, various communication interfaces in this domain have been hierarchical and unidirectional, although they can now normally be considered to operate in both directions, although electrical connections are beginning to do so. In the SG, the distribution domain communicates more closely with the real-time operations domain to manage energy flows related to the more dynamic market domain and other environmental and safety factors [61].

The distribution domain of a GXS refers to the part of the electrical network that is responsible for delivering electrical energy from transmission lines to end consumers, such as homes, businesses and institutions. It can include equipment such as transformers, substations, switches, smart meters and remote monitoring and control systems. Advanced communication technologies, such as wireless communication networks and standardized communication protocols, can also be used to facilitate communication and coordination between the various elements of the distribution network.

This field can be divided into several levels, such as low-voltage (LV) distribution, medium-voltage (MV) distribution and high-voltage (HV) distribution. Low-voltage distribution networks are responsible for delivering electrical energy to residential and commercial customers, while medium- and high-voltage distribution networks are used to supply electrical energy to large businesses and institutions.

SG in the distribution field uses state-of-the-art technologies to monitor, control and optimize the distribution of electrical energy in real time. Technologies such as sensors, communication systems and advanced analysis software are used to collect real-time data on energy consumption and network performance, and to make informed decisions on how to manage electricity distribution.

It ensures :

- Greater reliability.
- More efficient use of energy.
- Reduced energy losses.

- Improved energy quality.
- Greater capacity to integrate renewable energies.

- Consumer sector :

The distribution network supplies medium voltage to industrial consumers and low voltage to a very large number of consumers such as commercial and residential [55], [62], [63].

In a SG, consumers are key players in the system, as they are directly involved in energy management. SGs enable consumers to play an active role in managing their energy consumption, by providing them with real-time information on their consumption and giving them the opportunity to control their consumption.

This domain can include several parts, which interact with the SG, such as :

- Smart meters: to measure energy consumption in real time and communicate this data to the power grid. They also enable consumers to monitor their energy consumption in real time and make informed decisions about energy use.

- Smart appliances: are equipped with sensors and communication technologies to enable interaction with the power grid. They can be programmed to operate during off-peak hours and can be remotely controlled to optimize energy use.

- Energy storage solutions: enable consumers to store energy generated by renewable sources for later use. Energy storage systems can also be used to reduce energy demand during peak periods.

- Energy demand management programs: enable consumers to reduce their energy consumption during peak periods in exchange for financial incentives. These programs can be implemented by using smart technologies to control the energy consumption of appliances.

- Communication solutions: enable consumers to communicate with the power grid and receive information on electricity prices and energy demand management programs. Communication technologies can include mobile applications, websites and SMS notifications.

The consumer domain in a GXS consists of a set of technologies and solutions that enable consumers to make informed decisions about their energy consumption and interact with the power grid more intelligently and efficiently.

We can therefore deduce that the main characteristics of the consumer domain in an SG include :

- The ability to monitor and control energy consumption in real time.

- Integration of renewable energies and small-scale energy storage to optimize energy use.

- The possibility of selling surplus energy produced by consumers.

- Active consumer participation in energy demand management programs to reduce electricity consumption during peak periods.

- The ability to receive real-time information on electricity prices to make informed decisions on energy use.

- The use of intelligent technologies to automate tasks such as programming appliances to operate outside peak hours.

- • Service provider domain :

The smart grid service provider domain offers electricity services to customers and can communicate with the operations, markets and consumers domain [64], [65].

In a GMS, this domain is very vast and can include several types of services:

- Demand management: consumers can reduce their energy consumption during peak periods by receiving advice on how to use energy more efficiently.

- RE management: service providers can help consumers install solar panels and wind turbines to generate RE. They can also help consumers store the energy they produce for later use.

- Network optimization: network management software is used to monitor and optimize energy distribution. This can include detecting and correcting power failures, as well as reducing energy losses.

- Metering services: service providers can supply smart meters that enable consumers to track their energy consumption in real time. This can help consumers understand how they use energy and identify ways to reduce their consumption.

- Billing services: intelligent billing systems enable consumers to pay for the energy they use. These systems can include differentiated tariffs that encourage consumers to use energy during periods of low demand.

These examples are just a few of the many services that service providers can offer in the SG field.

Utilities will produce new and innovative products and services to meet the new needs and opportunities presented by the evolution of the SG [48].

- • Area of operation :

When we speak of the operating domain, we mean the control and command center, where we find most of the tasks related to energy movements, production centers, energy transport mechanisms and consumer homes. Network control centers make it possible to process the various tasks in real time and remotely [66].

Operations in a GXS are often managed by sophisticated software that enables automated decision-making in real time, to optimize energy flows in the network, improve energy efficiency and reduce costs.

The operation domain uses a two-way communication network to connect to local stations, customer networks and other intelligent field devices. It provides network operation, monitoring, control, fault management and important process decisions with the use of control and supervisory systems such as SCADA (Supervisory Control And Data Acquisition) [55].

The operations domain in a SG uses a variety of software to manage, monitor and optimize the smart grid. Here are a few examples of software commonly used in the operations area of a SG:

- ➢ SCADA (Supervisory Control and Data Acquisition): this is a control and supervision system that enables real-time monitoring of smart grid equipment and processes, such as power plants, transformers, power lines, smart meters, etc. SCADA is used to collect and analyze data, in order to detect problems, faults and anomalies and take the necessary steps to remedy them. SCADA is used to collect and analyze data, in order to detect problems, faults and anomalies and take the necessary steps to remedy them.

- ➢ EMS (Energy Management System): an energy management system for planning, coordinating and controlling the production, distribution and consumption of energy in the GMS. The EMS uses optimization algorithms to maximize energy efficiency, minimize costs and guarantee energy quality.

- ➢ Enterprise Asset Management (EAM) systems: EAMs are software applications used to manage EMS assets, including production equipment, distribution equipment and network infrastructure. EAMs enable equipment condition monitoring, maintenance planning, spare parts management and investment management.

- ➢ DMS (Distribution Management System): this is a distribution management system that monitors and controls energy distribution in the GMS, using advanced technologies such as sensors, PLCs and communication networks. The DMS is used to manage outages, overloads, power cuts and voltage imbalances.

- ➢ ADMS (Advanced Distribution Management System): this is a more advanced distribution management system that integrates SCADA, EMS and DMS functionalities to provide a complete overview of the smart grid. ADMS uses advanced analysis tools, such as load modeling, demand forecasting and scenario simulation, to make informed decisions in real time.

- ➢ OMS (Outage Management System) : this is an outage management system that detects, locates and resolves outages in the GXS. OMS uses mapping, communication and coordination tools to minimize the impact of outages on consumers and energy suppliers.

- • Marketing :

A new market is developing with the advent of EMS. There are new services to support and new products to sell, which affects the flow of energy. A new market emerges, based on energy trading and the provision of ancillary services [67].

Smart grid marketing focuses on the commercialization of technologies and services associated with optimized energy management.

Companies developing EMS technologies need to promote their solutions to utilities, regulators, consumers and other key stakeholders. Marketing strategies can include participation in industry events, publication of white papers and other research documents, as well as promotion online and in traditional media.

Marketing companies can help energy suppliers communicate with customers about how GXS can benefit them.

This domain contains a two-system architecture [68] :

- ➢ Systems linked to Commercial Virtual Plant Management (CVPM), providing market access to an aggregation of decentralized generation plants and flexible consumption resources (aggregation of flexible consumption points).

➢ Technical Virtual Plant Management (TVPM) systems, ensuring network stability and optimizing energy networks as a whole, by taking into account the technical model of all the generation, consumption and storage resources provided by TVPMs, as well as the technical constraints and limitations of network infrastructure management.

3.3.   The objectives of a smart grid

The objectives of an SG are numerous, and can vary according to the needs and priorities of each country or region. However, here are some of the most common objectives associated with smart grids:

- Integration of renewable energy :

SGs enable the harmonious integration of renewable energies into the power grid, effectively managing the production and distribution of electricity, which can be variable and intermittent. This helps to reduce dependence on fossil fuels and cut greenhouse gas emissions.

Today, we're no longer talking about the old centralized production systems, but about decentralized systems, which are much more interesting, since they guarantee sufficient operational flexibility.

The information system that forms part of smart grids can detect production and consumption rates not only in the short term, but also in the long term.

In fact, today's distribution networks are balanced by the interaction between thousands of control points during decentralized injection, unlike traditional networks which only guaranteed this balance between supply and demand through a few hundred of these points [69].

- Improving power system reliability :

SGs coordinate network components autonomously, so they're not just used to control production facilities, but also to connect or disconnect loads according to consumer needs.

They can reduce power outages by monitoring and controlling the power network more effectively. This also enables faster response in the event of disturbances or incidents on the network.

The presence of high-speed protection equipment is essential to manage energy flows, isolating faulty sections and giving remote control the advantage of reconfiguring certain zones based on incidents that have occurred and even planned interventions.

Real-time interaction with measuring instruments to help balance the network by synchronizing energy supply and demand. For this reason, each transformer station must contain control, protection and automation equipment. The latter, applied to the distribution network, is needed to interact bidirectionally with the decentralized producer, while modernizing new control-command approaches based on new visualization techniques for decision-making.

- Energy optimization :

Energy optimization in a smart grid is a process that aims to maximize the grid's energy efficiency while minimizing costs and reducing greenhouse gas emissions.

Several methods can be used to optimize energy, including :

- Demand management.
- Production management.
- Energy storage.
- Distributed energy.
- Dynamic pricing.
- Data analysis.

By using these methods, a GMS can reduce energy production costs, improve energy efficiency and reduce greenhouse gas emissions, thus contributing to a more sustainable energy economy.

- • Managing individual consumption :

Users can manage their energy consumption throughout the day, while grid operators can detect potential faults more quickly. As mentioned above, sensors interact with each other and are the first applications of the smart grid, playing a key role in its development. They give consumers access to information on electricity prices, consumption quality and peak times. Representing the first building block of the SG, these sensors are linked to good energy management and the birth of competition.

- • Reducing energy costs:

By encouraging users to reduce their energy consumption at times when demand is high. This can be achieved by offering different tariffs depending on the time of day, or by using financial incentives.

Smart grids enable two-way communication between energy producers and consumers. This means that users can not only consume electricity, but also generate it and sell it to the grid.

EESs optimize the use of existing resources and infrastructure, reducing operating costs for electricity suppliers and consumers alike.

SGs enable value-added services such as real-time monitoring of energy consumption, energy analysis tools and energy cost information.

- • Integrating new technologies :

Urban infrastructure will undergo a major change when electric vehicles appear on the scene, affecting the power grid as well. The development of these vehicles will only be achieved when the necessary charging spaces are provided (shopping centers, safe charging structures, in homes, streets, etc.).

Electric vehicles can be charged according to the capacity of the electrical grid. Owners can program their cars to charge during periods of low demand, when the grid is less busy.

These vehicles can be used as a source of back-up power, with the batteries acting as an emergency power source in the event of a power cut. Electric vehicle owners can sell their stored energy to power grids at peak times.

3.4. Research on smart grids

Researchers Y. Cunjiang et al (2012) [70] have carried out an analysis of the current network and highlighted its drawbacks, such as energy loss, vulnerability to outages, high costs, greenhouse gas emissions, capacity limitations and so on.

In line with the current demands of energy users, an evolution of current power grids is underway, aimed at resolving some of the drawbacks listed above. Smart grids use information and communication technologies to monitor, control and optimize the production, distribution and consumption of electricity. The researchers also presented the characteristics of the smart grid and its technical components, to finally propose a smart grid architecture.

Researchers S. Goel et al (2013) [71] have drawn up a vision of SG by 2030 based on a perspective paper. They discussed the imperatives of the various players in this enterprise and the obstacles to realizing this vision. They began with a discussion of how the power grid works today, including the role of generation, transmission and distribution. They then discussed the problems with this network and the obstacles to achieving the smart grid communications vision. This debate was followed by the SG's objectives.

Researcher M.E. El Hawary (2014) [72] presented, in his paper the SG and associated technical, environmental and socio-economic benefits, as well as other non-tangible benefits for society, and explained the need for the concept and the fact that it is a dynamic, real-time interactive infrastructure that meets the challenges of designing and building the power system of the future, rather than simply being a marketing term. To illustrate the diversity of terminology, the author compared a definition from the Electric Power Research Institute (EPRI) with that proposed by a study group of the International Electrotechnical Commission (IEC). He then cited three examples of definitions to highlight the diversity of views on the smart grid. Initial misconceptions and characterizations of the smart grid are discussed as a prelude to solving the challenging issues driving the development and implementation of related innovative technologies, products and services. Then, the researcher addressed some of the often-cited barriers to acceptance of the SG that are based on concerns and issues related to its evolution, adoption and acceptance.

Researchers M.L. Tuballa et al. (2016) [73] presented an overview of SG with its general characteristics and functionalities. They discussed fundamental and related SG technologies and identified research activities, challenges and issues. They showed how these technologies have shaped the modern power grid and have continued to evolve and enhance its role in better aligning energy demand and supply. Smart grid implementation and practices in various locations are also revealed.

Researchers S. Howell et al (2017) [74] have developed this paper on the concept of smart grids by critically reviewing the literature. This includes a discussion of modern concepts such as "smart grid", "microgrid", "virtual power plant" and "multi-energy system", and the relationships between them, as well as trends towards distributed intelligence and interoperability. Each of these emerging urban energy concepts has merit when applied in a centralized grid paradigm, but very little research applies these approaches in the energy landscape characterized by high penetration of distributed energy resources, prosumers (consumers and producers), interoperability and big data.

As these fields continue to expand, this will lead to new challenges and opportunities as the status of energy systems changes radically. A new generation of holonic energy systems is needed to orchestrate the interaction between these dense, diverse and distributed energy components. These researchers have therefore contributed to a description of holonic energy systems and the implicit research needed for sustainability and resilience in the impending energy landscape.

Researchers S. Kakran et al. (2018) [75] have provided a detailed description of advances in demand-side management, demand response programs, distributed generation, the technical issues involved in their progress and the key benefits. Indeed, renewable energy resources are also becoming an important part of distributed generation, providing a solution to the environmental problems caused by conventional power plants. Few countries are working to deploy advanced metering. At the same time, the scope of research into various smart grid technology programs has been explored.

Researchers Y. Ming et al. (2019) [76] proposed an efficient privacy-preserving multidimensional data assembly scheme in a smart grid by virtue of homomorphic encryption and super increasing sequence. Security analysis indicates that the proposed scheme proved to be secure in the random oracle model, satisfying all security and privacy requirements. The in-depth performance analysis shows that, compared with related schemes, the proposed scheme achieves the lowest computation and communication costs, thus suitable for practical applications.

Researchers D. Tang et al. (2019) [77] modeled a smart grid coupled with a social network and studied its vulnerability to false pricing attacks in the social network. The energy consumption profile based on social information is modeled as a consumption reordering problem, which aims to maximize the benefit of demand management. The process of spreading false prices is described by a multi-level influence propagation model, which takes into account the personality of end-users. Different attack strategies are considered and the response of the power operator is modeled. Residual distribution line amperage and expected unsupplied energy are adopted to quantify the impact of attacks on the power system. To account for the stochastic characteristics of the influence propagation process, a Monte Carlo simulation is used. The proposed modeling and analysis framework is applied on a modified IEEE 13-node test loader and a fictitious social network. This promotes the systemic characteristics of autonomy, belonging, connectivity, diversity and emergence, and balances global and local system goals, thanks to adaptive control topologies and demand-responsive energy management.

4. Conclusion

It is important to note that increasing the share of renewable energies can have adverse consequences for the stability and smooth operation of the power system as a whole. Current generators, based on renewable energy sources (RES), cannot provide system services, and their output may be difficult to predict. For this reason, the transition to smart grids is essential. Like any complex system, a smart grid is made up of a large number of interacting entities. This dynamic entanglement means that a rigorous modeling process is essential.

<u>Chapter 3:</u>

Presentation of modeling methods

<u>Chapter 3:</u> Overview of modeling methods

1. Introduction

Several approaches, such as systems engineering, have proposed methodological approaches associated with tools providing simplified representations of reality. This is in order to master the understanding, design, development and operation of complex systems [79].

Modeling an intelligent power network as a complex system enables us to understand the interactions between the various network elements, anticipate unforeseen events and make more informed decisions about network planning and operation.

Our third chapter will provide a detailed classification of the different modeling methods, exploring their fundamental principles, their specific applications, as well as their advantages and limitations.

2. Modeling methods

SGs are among the most complex of systems, and this complexity has led to the emergence of the modeling process. Indeed, the modeling and design of this decentralized, autonomous and distributed system represents one of the greatest challenges of the $21^{\text{ème}}$ century. Modeling the network, user behavior, reliability and realism is crucial to preparing the way for future technologies.

Modeling, in general, involves developing a set of equations or rules to describe a system in a reproducible and simulable way. Its main objective is to control complexity.

The modeling process is a well-structured methodological approach aimed at representing and explaining the qualitative data of a system and its interactions internally and with its external environment, by developing conceptual modeling methods or languages to achieve a clearly identified goal [79].

The application of this process therefore requires the use of certain suitable modeling methods. These modeling methods are generally classified and grouped into five families: functional modeling, decision modeling, resource modeling, information modeling and mixed modeling. Indeed, for the analysis and modeling of complex systems, there are different categories of classical methods, as well as a multitude of application domains [80].

2.1. Functional modeling

Function-oriented modeling consists in describing the functions, activities and designed process of a company. Functional modeling methods represent the interactions between functions and activities by describing the information exchanged between them and the resources used, and decompose them in an organized and detailed way to better understand how a business operates [81].

Various methods have been used to analyze and model functions, such as SADT, the IDEF family and Petri nets.

2.1.1. The SADT method

SADT (Structured Analysis and Design Technique), also known as IDEF0 (Integration Definition for Function modeling), is a method developed by Doug Ross in 1977 in the USA, and introduced to Europe in 1982 by Michel Galiner. In the 1980s, this method became a standard for the graphical description of complex systems, thanks to its top-down approach to functional analysis. SADT/IDEF0 is a graphical method for describing complex systems comprising various flows of material and work, whether automated, servo-controlled or incorporating IT aspects. It adopts a general-to-particular approach for a structured and comprehensive analysis of systems.

A system is described in the form of a coherent sequence of diagrams. The highest-level diagram represents the purpose of the technical system. Each lower-level diagram defines the system's sub-functions, as well as their relationships and arrangements within the system. By convention, the highest level is referred to as A-0. This A-0 level is broken down into n boxes A1, A2, ...An [82].

2.1.2. The IDEF method

IDEF represents the series of functional modeling methods developed by the US Department of Defense from the 1960s onwards: IDEF0 (activity-based modeling), IDEF2 (models for simulation) and IDEF3 (capture of process descriptions) [83].

The IDEF0 method was developed with reference to the SADT method to analyze and communicate the functional aspect of a system and describe processes in the form of a hierarchical model [84]. It is used to model the activities and their interrelations of an organization or system [85]. These activities can be decisions made, information communicated or resources used [81].

The IDEF2 method can be used to model a system operating with queues. This method responds to the shortcomings of the SADT/IDEF0 method in terms of dynamic analysis of such a system. Indeed, it consists in modeling the physical system and its control, entity flows and resource management [86].

IDEF3 fills in the gaps left by IDEF0 in terms of modeling corporate control flows [87]. It consists in graphically describing the processes or activities and transitions of objects used in a system [84]. However, it does not manage resources and material flows, and is oriented towards modeling processes according to a well-chained methodology, with an emphasis on logical links [86].

2.1.3. Petri dishes

Petri nets can be used to model the operation and dysfunction of a complex system in a precise and rigorous way [88]. In fact, they can replace both functional and dysfunctional modeling methods. They are based on state-transition models [89]. These networks graphically represent the evolution of a production system. They describe the state of products before and after the execution of each task.

Researchers R. Abbou et al (2003) [90] have used the Petri net model to model a dynamic production line, integrating maintenance policies. The networks presented can cope with problems of concurrency, synchronism, parallelism and resource sharing, and handle a wide range of data relating to the production system [91].

On the other hand, Petri nets will be unreadable when modeling a complex system, and will be misunderstood by inexperienced people [88].

2.2. Business intelligence modeling

The aim of this approach is to provide a detailed description of the decisions to be made within a well-defined time horizon and in relation to activities. In fact, the decision is an interface between strategy and operation in the system [92].

The best-known decision-oriented modeling methods are GIM and its origin GRAI, originally developed in the GRAI research laboratory at the University of Bordeaux in the early 1980s [93].

2.2.1. The GRAI method

The GRAI (Graphes à Résultats et Activités Inter reliés) method is a conceptual model used as a reference in the field of production systems. It integrates the principles of systems thinking with the theories of discrete activities and production management.

The decision-making system is responsible for managing decisions to control a physical system, and these two systems interact with a third system, the information system, for communication.

The decision system is decomposed along two axes, a vertical axis linked to the nature of the decisions and a horizontal axis linked to the type of decisions [94].

The GRAI grid makes it possible to distinguish functional dependency connections (represented by a double arrow, implying the transmission of an instruction or objective) from informational connections (represented by a single arrow, implying the transmission of an information flow) between decision centers.

The main aim of this modeling is to ensure that activities at each level are synchronized, and that decisions are coordinated between the different levels.

2.2.2. The GIM method

GIM (GRAI Integrated Methodology) is an enterprise modeling method based mainly on three methods: GRAI, IDEF0 and Merise. The main tools and formalisms used in GIM are IDEF0 for the functional view and the physical model, GRAI grids and networks for the decision-making model, and the entity-relationship formalism for the information model [95].

GIM addresses four distinct perspectives:

- The Information view, for data and knowledge.
- The Decision view, which focuses on activity chains and decision centers.
- The Physical view, which covers material resources.
- The Function view, which deals with functional decomposition.

GIM proposes a three-tiered approach:

- The Conceptual level, which aims to define needs as perceived by users.
- The Structural level, which involves defining a technological solution that has been validated by users.
- The Realization level, which describes the implementation, taking into account organizational aspects, information technologies and industrial and manufacturing technologies.

GIM does not have a specific modeling language, but relies on existing formalisms such as Merise, GRAI networks, IDEF0, IDEF3 and the relational model. In addition, GIM adopts a two-part methodological approach, comprising a conceptual part and a technological part, to guide its interventions.

2.3. Resource modeling

Resource-oriented modeling methods allow the description of the resources needed to carry out an activity, taking into account the constraints on resource allocation. They are specialized in the management of resources from their acquisition to their exploitation, but without understanding how they work [96].

Resource modeling methods include the multi-agent approach, the PERA methodology and the MOVES language.

2.3.1. The multi-agent approach

The multi-agent approach aims to schedule and manage a complex industrial system by following a resource modeling or function design approach. This method is close to a systemic analysis approach. It allows for system customization and decomposition into interrelated entities considered as physical agents (machines, tools, etc.) and functional agents (inventory management, production management, maintenance management, etc.). These agents are able to control their reactions and interactions according to their states and purposes [87].

They can be competent physical or virtual entities, communicating with each other and interacting with their environment. They can reproduce themselves, using their own resources and providing services. This methodological approach is based on the following phases: analysis, specification, design, implementation and operation. It enables the system to be broken down into decision-making, information and operational sub-systems, while maintaining their interrelationships and identifying the necessary resources [83].

2.3.2. PERA method

PERA is an engineering methodology for industrial environments, developed by Prof. Williams. The architecture is broken down into five stages [97]:

- A conceptualization phase consisting of identification (defining the scope of the study) and conception (describing the company's mission, vision and values).

- A definition phase, during which implementation requirements are defined, along with the tasks, modules and macro-functions needed to meet them, and finally the connection diagrams between tasks, modules and macro-functions.

- A design phase which is broken down into a functional design phase (specification of the initial choices of information system architecture, human organization and operational part) and a detailed design phase (description of the details corresponding to the information provided in the functional design).

- An installation and construction phase, which consists of implementing the decisions taken during the previous phases in terms of installing and testing databases and programs, training staff, and installing equipment.

- An operational and maintenance phase, corresponding to the actual installation and upgrading of the system.

2.3.3. The MOVES language

Like MECI (Modélisation d'Entreprise pour la Conception Intégrée, or Enterprise Modeling for Integrated Design, built as part of the AICOSCOP project, which proposed a method to aid the design of production systems), the MOVES modeling language consists in representing a company's human resources based on their skills and their implications in the analysis of company performance. In fact, it enables the hierarchical description of the individual or collective skills of the company's players, who are capable of carrying out the full range of activities under well-studied environmental conditions, with a view to certifying the company's organizational and performance unit. However, this technique is mainly oriented towards defining the concepts to be used to model systems [97].

2.4. Information modeling

Information architecture is a combination of fixed structures and objects with short life cycles [98].

Information modeling methods are designed to model and process the company's information system, as shown by the OLYMPIOS method, unified languages, etc. They ensure the flow of information concerning processes, functions, resources and organization within the company. They ensure the circulation of information concerning processes, functions, resources and organization within the company. Certain languages have been developed to meet this need, such as UML, UEML and IEM.

2.4.1. The UML language

UML (Unified Modified Language) is a fusion of the Booch, OMT (Object Modelling Technique) and OOSE (Object Oriented Software Engineering) methods [99].

UML has gained widespread acceptance thanks to a general consensus. Many industrial players have adopted UML and are contributing to its evolution. In a short space of time, it has become an essential standard.

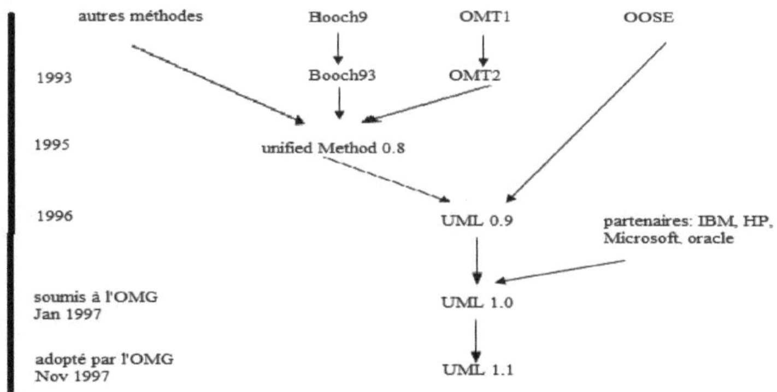

Figure 14UML creation history

UML is a visual language comprising a collection of diagrams offering different points of view on the project in question. With UML, we have diagrams to represent the software to be developed, including how it works, how it is deployed, what actions it can perform, etc.

The diagrams of a UML language are hierarchically dependent and complement each other, so that a complete project can be modeled and analyzed. Since UML 2.3, there have been fourteen such diagrams. The following figure shows the hierarchical location of a UML diagram [100].

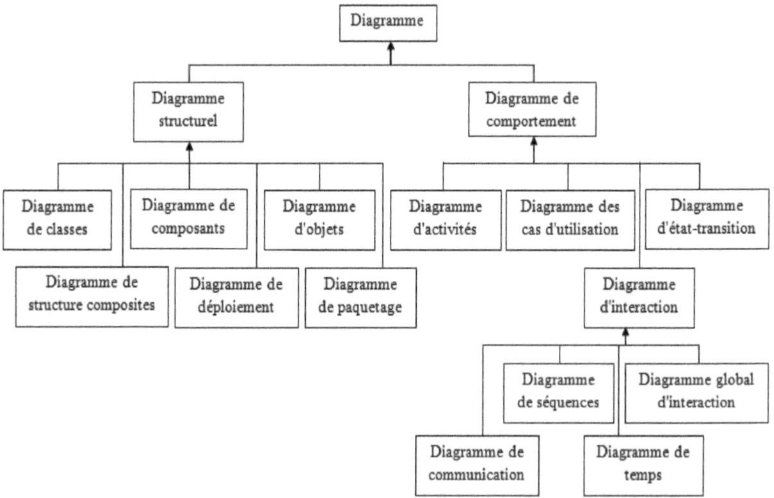

Figure 15Hierarchical location of a UML diagram [91].

Understanding UML can be complex because it has several facets:

- It is a standard.
- It is an object modeling language.
- It serves as a means of communication.
- It provides a methodological framework.

The elements or diagrams of the language are not defined on mathematical grounds, which makes it impossible to check the semantic consistency of a UML diagram mathematically. What's more, we may come across two different types of diagram, to describe the same system from two different views. It is impossible to formally verify the consistency of the overall model [101].

Furthermore, experience has shown that certain weaknesses of UML need to be overcome to make it an effective language for engineers [102]. These include :

- It is necessary to incorporate requirements into the UML model and guarantee their traceability to the design stage.

- Need to represent non-software elements and specify their type and discipline.

- Need to represent performance and physical attributes.

- Need to specify data types for physical elements.

- Need to show explicit modeling elements to represent physical inputs and outputs.

- Need to improve semantics for specifying events and linking the event with the action.

Returning to the gaps, SysML (Systems Modeling Language) was born as an extension of the object-oriented UML language to model all stages of complex, multidisciplinary system design. SysML mainly solves the shortcomings of other profiles if it is used in the upstream phases of systems engineering. In addition, traceability is a design requirement targeted by SysML [91].

2.4.2. UEML language

UEML is a thematic network project funded by the European Union and has been developed to ensure the exchange of information, data and knowledge between the various standardization bodies specializing in business modeling and integration. This model analyzes the most widely used operational languages, but is used in coordination with operational methods for general enterprise design [103].

The development of UEML must take into account the following principles:

- The language will be defined as a finite set of constructions.

- Principle of separation of processes and resources.

- Principle of separation of corporate behavior and corporate functionality.

- Principle of separation of resources and organizational units.

It's important to clarify one point: the name UEML may evoke UML, but the two should not be confused. UML is a general modeling language mainly used in information systems, but it can be applied more broadly. UEML, on the other hand, is a specialized language specifically dedicated to business modeling. It is based on concepts familiar to business users.

2.4.3. The IEM method

IEM is a generic object-oriented modeling method [84] for describing the information and functions of objects in a manufacturing system, with reference to the IDEF0 and CIMOSA models [101]. Unlike the IDEF0 method, IEM presents the three states of an activity's inputs and outputs: order, product and resources [87].

The IEM technique can be used to model discrete processes and to study the interactions between resources and control elements for the execution of activities [84]. It makes it easy to organize processes, record data and facilitate the production and circulation of information [104].

On the other hand, it does not allow detailed interactions to be represented for complex discrete systems. It has shortcomings in terms of problem communication between system designers and simulators [84].

2.5. Mixed modeling

A mixed-method approach is one in which the researcher integrates both quantitative and qualitative data and methods into a single study.

This modeling technique studies organization, resources and information process models, and quantitatively analyzes the functioning of a company and its production system [81].

Although they cover several aspects and study various systems (functional, informational, resource and organizational) of a company, mixed modeling methods cannot model decision-making systems.

2.5.1. The CIMOSA method

The CIMOSA method, developed by the AMICE Consortium as part of the ESPRIT project [83], models the company's different domains (resources, organization, function, information) and designs an overall production system in a more relevant way (Figure 26) [104]. In fact, it integrates two incoherent systems and makes them homogeneous by semantically unifying the concepts of the different systems [105]. It is the most widely used modeling method, representing the various views [106] and resources essential for process execution [87]. However, this method does not schematize activities and processes, and is not directly adapted to simulation [84]. To this end, it describes them using a formal language for integrated modeling of different aspects of the business [86]. CIMOSA has been used to develop certain modeling methods, such as ARIS, which is based on software engineering and the conceptual and organizational modeling of integrated systems, and CENENV40003, which more precisely formulates the fundamental principles for enterprise modeling [107].

The architecture in question comprises a modeling framework, an integration platform and an intervention methodology. The modeling framework, known as the "CIMOSA cube", is based on three fundamental and independent principles, represented by the three axes of the cube.

> The genericity axis can be broken down into three distinct levels:
> - The generic level, which defines the basic elements of the modeling language, called "constructs".
>
> - The partial level, which offers predefined, reusable structures for specific fields of application.
>
> - The special level, which encompasses company-specific models.

The generic and partial levels form CIMOSA's reference architecture, subject to possible standardization.

> The model axis, also known as the derivation axis, proposes a classification into three modeling levels:
> - The needs definition level, which consists of drawing up the specifications.
>
> - The design specification level, which involves a conceptual analysis of solutions in response to expressed needs.
>
> - The level of description of the implementation, which requires a detailed description of the chosen solution.
>
> The view axis, also known as the generation axis, defines the company from four distinct perspectives:

- The functions view, which concerns the company's functionalities and behavior in terms of processes, activities and operations.

- The information view, which describes the company's information system.

- The resources view, which covers the means required to implement the company's functions.

- The organization view, which describes the distribution of responsibilities and authority in the decision-making process.

Each view offered by the CIMOSA modeling framework is not an isolated sub-model, but rather a distinct filter for interpreting the information contained in the model.

2.5.2. The GERAM method

Based on the well-known CIMOSA, GRAI/GIM and PERA models, the GERAM meta-method has developed, in a more relevant way, organized and interrelated concepts for integrated enterprise modeling [103]. It is an enterprise engineering methodology based on the PERA and GIM methods, it presents the life cycle of a system based on PERA and describes the different views of the enterprise using aspects of the CIMOSA method [104]. GERAM has seven interrelated methodological elements admitting a methodology (EEM), a language (EML), a framework (CEMC), a model (PEM and EM), tools (EET) and modules (EMO) for enterprise engineering [84].

Based on the GERAM method, researcher John Zachman developed the Zachman method , which presents a general architecture framework. However, although GERAM contributes to the unification of two distinct approaches to enterprise integration: product-oriented and process-oriented approaches, it does not advocate the use of specific methods or tools to model all aspects of the enterprise [108].

3. Conclusion

Once we have thoroughly explored the complexity and interconnectedness of a smart grid through in-depth modeling, the next step will be to optimize energy management within this system. Indeed, the ultimate goal of these new-generation infrastructures is to guarantee efficient, reliable and sustainable electricity distribution. The ability to dynamically control energy flows, balance production and consumption, anticipate peaks in demand and harmoniously integrate renewable energies will be decisive.

Chapter 4:
Optimal Energy Management Techniques

<u>Chapter 4:</u> Optimal energy management techniques

1. Introduction

As the power output of renewable sources is intermittent and dependent on a number of uncontrollable conditions, an effective management system is needed to make decisions for better energy utilization. An optimal power management strategy should result in an efficient, reliable and cost-effective system [109].

The question of economic efficiency is a major concern when planning RE projects. Low economic efficiency is one of the main arguments put forward against RE sources. A solution with the best economic benefits is generally the one that is implemented. Technical or ecological aspects are often considered of secondary importance [5].

The components of a multi-source system must meet several requirements:

- Satisfy the electricity needs of residential homes.

- Optimize energy costs over the entire life cycle.

- Minimize overall system costs.

To achieve these objectives, the user needs to have data on load demand, RE resources over several years, as well as economic and technical data [110].

For this reason, sizing is a necessary phase for any HS in order to identify the power and capacity required for the various sources and other energy conversion devices and to best meet the system's requirements, while taking into account the location where these are installed and the costs of building the system according to a worst-case mission profile [111].

Energy sizing and optimization are closely linked in the sense that the sizing of an energy system must be optimized to minimize energy consumption and maximize energy efficiency.

Engineers have used optimization methods to optimize the sizing design of HS devices, their components and sometimes even their control systems for decades [52].

In what follows, we discuss the main sizing tools described in the literature, as well as the methods commonly used for optimization in the field of HS sizing and their operations associated with smart grids.

2. The main sizing tools

There are several software packages for sizing and simulating hybrid energy systems [112], including HOMER, SOMES, Hybrid2, RAPSIM, PVsyst, INSEL and others. All these software packages can help determine the optimal size of each component of a HS according to several parameters such as energy demand, availability of energy sources, system reliability and available budget, but their optimization strategies are different.

2.1. HOMER

HOMER (Hybrid Optimization of Multiple Energy Resources), developed by the National Renewable Energy Laboratory (NREL), is a software package for modeling and sizing hybrid renewable energy systems. It is a powerful tool, thanks to its ability to model different RE sources, use optimization algorithms to determine optimal system size and provide detailed reports on system performance [113].

Its many strengths include

- It can model a wide range of renewable energy sources, including solar, wind, hydro, biomass and hydrogen. It also takes into account diesel generators and batteries to provide an energy storage solution.

- It carries out a system reliability analysis to identify risks of failure and determine the safeguards needed to guarantee system reliability.

- A user-friendly interface makes it easy to model the SH, adjust parameters and visualize results.

- Detailed reporting on system performance, including costs, energy production, energy demand, battery state-of-charge, CO_2 emissions and optimization results.

2.2. Hybrid2

The Hybrid2 software package is a user-friendly tool for detailed long-term performance and economic analysis of a wide variety of hybrid power systems. Hybrid2 is a probabilistic computer model, using time-series data for loads, wind speed, solar insolation, temperature and the power system designed or selected by the user, to predict hybrid power system performance [114]. Variations in wind speed and load at each time step are taken into account in the performance predictions. The code does not take into account short-term system fluctuations caused by system dynamics or component transients.

Hybrid2 allows the user to :

- Carry out a detailed analysis of potential systems, taking into account a wide variety of inputs ranging from taxes to load information.

- Have an easy-to-use user interface, so they can visualize simulation results, adjust parameters and generate reports.

- Analyze the impact of different energy use scenarios.

2.3. SUMES

Developed by Utrecht University in the Netherlands, SOMES (System Optimization for Multi-Energy Systems) is a simulation and optimization software package for multi-energy systems.

SOMES models and optimizes complex energy systems, taking into account multiple energy sources such as renewable energies (solar, wind, hydro), energy storage systems, conventional generators (thermal, diesel, gas) and electricity distribution networks.

The simulation is carried out on an hourly basis. The optimal system is sought by comparing the costs of several systems, within user-defined limits [20].

Its main features are :

- Modeling and simulation of multiple energy sources for a given energy system. Meteorological data, geographical parameters and user load profiles are taken into account for simulation.

- The use of optimization algorithms to find the optimum combination of each energy system component, such as the size and capacity of power generators, storage systems and distribution networks.

- Formulation of a complete economic analysis of energy system costs, including equipment, maintenance and operating costs.

2.4. RAPSIM

RAPSIM (Renewable Energy Alternative Production Simulation), developed by the Murdoch University Energy Research Institute in Australia, is a sizing and simulation software package for renewable energy systems.

Optimum sizing is achieved by trial and error. Users change system parameters (number of batteries, wind turbines, diesel generator power) and judge the result to choose the best solution for their needs [20].

The main features of RAPSIM are as follows:

- The ability to model different types of components for RE systems, including solar panels, wind turbines, diesel generators, batteries and inverters.

- The use of simulation algorithms to model the behavior of each component of the RE system and to estimate energy production.

- Sizing of the various components of the renewable energy system, taking into account the system's energy requirements and environmental constraints.

- RAPSIM provides a complete economic analysis of the RE system, including equipment, maintenance and operating costs.

2.5. PVsyst

PVsyst (Photovoltaic System Analysis Software) is a solar panel sizing software that provides information on energy production, irradiation, installation costs, surface area required and annual energy production. An advanced mode allows much more information to be obtained for a very comprehensive study [115]. Thus, for one or more grid-connected or stand-alone installations, PVsyst can be used to obtain the optimum configuration after a technico-economic study.

It offers the possibility of :

- Provide a complete economic analysis of PV systems, including installation, maintenance and operating costs, as well as the savings achieved through the use of solar energy.
- Integrate meteorological data for the installation site from online sources such as Meteonorm, or from data measured on site.
- Use advanced mathematical models to simulate energy production from PV systems as a function of weather conditions, site characteristics and system parameters.

2.6. INSEL

INSEL (Integrated Simulation Environment Language), developed by the University of Oldenburg in Germany, is a simulation and sizing software package for renewable energy systems, including photovoltaics, wind turbines and energy storage systems. It allows users to create a structure using its library with a specified runtime [116]. This simulation software has the flexibility to create system models and configurations for planning and monitoring electrical and thermal energy systems.

Its advantages are :

- It provides a complete economic analysis of the RE system, including the costs of equipment, maintenance and operation, as well as the savings achieved through the use of RE.

- It has its own database of meteorological parameters from almost 2000 locations worldwide, photovoltaic systems, thermal systems and other devices.

- It uses simulation algorithms to model the behavior of each component of the RE system and to estimate energy production.

3. Energy optimization methods

Optimization is the search for the best among a large number of possible solutions. The main objective of optimization in the cost-effectiveness of SHs, including production sources and even its associated components, is to determine a set of parameters, topology and number of units used to satisfy certain specifications and objectives under design constraints, such as power output, efficiency, volume and cost [52].

There are many energy optimization methods, each with its own advantages and disadvantages, and they can be classified into two main families: deterministic and stochastic methods, as shown in figure 28.

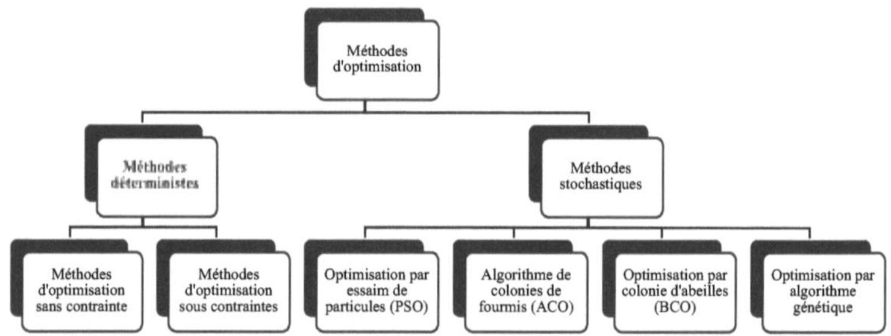

Figure 16Classification of some optimization methods

3.2. Deterministic methods

Deterministic optimization, or optimization by mathematical programming, embodies algorithms that rely heavily on linear algebra, as they are generally based on gradient calculus. Deterministic optimization has both advantages and disadvantages. The most notable advantage is that convergence to a solution is much faster than optimization by stochastic algorithms [117].

They have the advantage of providing precise, deterministic results that can be easily interpreted.

Its limitations can be seen in its dependence on the initial solution, its sensitivity to errors or inaccuracies in the input data, and the limitations of the mathematical models used, which may be limited in their ability to represent reality accurately.

There are two main aspects to deterministic optimization: unconstrained optimization and constrained optimization [52].

3.2.1. Unconstrained optimization methods

Unconstrained deterministic optimization methods seek to find the optimal value of an objective function without taking constraints into account. These methods are often used to optimize continuous, differentiable and unimodal functions [118], [119].

We cite the following approaches:

- Newton's method.
- The gradient descent method.
- The quasi-Newton method.
- The Levenberg-Marquardt method.

These deterministic unconstrained optimization methods are often simple and effective for solving relatively straightforward problems. However, they can be limited by their ability to converge rapidly to an optimal solution, particularly for objective functions that are non-convex or have several local minima.

3.2.2. Constrained optimization methods

Constrained deterministic optimization methods are used to find the best possible solution to a problem while satisfying certain constraints.

Here are a few examples of constraint-based optimization approaches:

- Logarithmic barrier method.
- Sequential quadratic programming method.
- Confidence region method.
- The Lagrangian method.

3.3. Stochastic methods

Stochastic optimization includes optimization methods in which chance is present in the search procedure. This is a fairly general definition for stochastic optimization, in which chance can be included in many ways [117].

Stochastic optimization methods are used to solve optimization problems in which the data is random or imprecise. These methods are often used for high-dimensional problems, or for problems that cannot be solved efficiently using deterministic methods.

3.3.1. Genetic algorithm optimization

The genetic algorithm is an optimization technique based on the principle of natural selection for reproduction and for various other operations such as crossover and mutation [120].

Genetic algorithms are optimization methods. They take their name from the biological evolution of living beings in the real world. They seek to simulate the process of natural selection in an unfavorable environment, inspired by C. Darwin's theory of evolution.

Such that the best-adapted "individuals" tend to live long enough to reproduce, while the weakest tend to disappear [121].

It works on the principle of selecting the best and discarding the rest. This selection combines a strategy of survival of the fittest with a random but structured exchange of information [122].

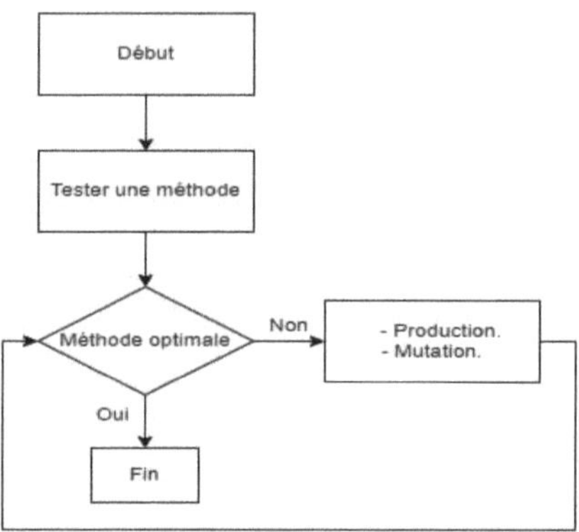

Figure 17Process of finding an optimal method using a genetic algorithm

A genetic algorithm seeks to find the optimal points of a function defined on a data space. To use it, five elements are required [123]:

> The principle of coding the population element: this generally follows a phase of mathematical modelling of the problem being addressed. The success of genetic algorithms depends on the quality of data coding.

> A mechanism for generating the initial population. This mechanism must be capable of producing a non-homogeneous population of individuals that will serve as the basis for future generations.

> A function to be optimized which returns a value of R+ called the "objective" function.

> Operators to diversify the population over generations and explore state space (crossover and mutation).

> Sizing parameters: population size, total number of generations or stopping criterion, application probabilities for crossover and mutation operators.

3.3.2. Particle Swarm Optimization (PSO)

Particle Swarm Optimization (PSO) is a heuristic optimization method inspired by the behavior of bird or insect swarms to solve optimization problems. PSO is based on the simulation of a swarm of particles moving through a search space, exploring different possible solutions.

When birds flock collectively together, they present a beautiful, coordinated pattern. They follow a behavioral pattern in which each agent follows three rules [124]:

> Separation: each agent tries to move away from its neighbor if both are too close in relation to the separation between neighboring agents.

- ➢ Alignment: they line up either in a straight line or in a parallel formation.

- ➢ Cohesion: each agent tries to position itself in the middle between its neighbors.

The objective is to find the best possible solution by minimizing or maximizing a given objective function. Key PSO parameters include particle number, fitness function, maximum permissible speed, inertia factor and acceleration coefficients.

3.3.3. Ant colony algorithm

The Ant Colony Algorithm (ACA) is inspired by the foraging behavior of ant colonies. Ants follow pheromone trails left by other ants to find food sources and bring food back to the colony.

The longer the path, the smaller the pheromone deposition, and the shorter the path, the higher the pheromone deposition [125].

Possible solutions are represented by paths in a graph, where each node represents a step in the solution and each edge represents the transition between two steps.

Artificial ants are generated to traverse the graph, following the pheromone trails left by other ants. The probability of an ant following a particular edge is a function of the amount of pheromone deposited on that edge and the distance between nodes [126].

Key ACA parameters include the amount of pheromone deposited by the ants, the rate of pheromone evaporation, the ant's decision factor and the edge visibility factor.

3.3.4. Bee colony optimization (BCO)

Bee Colony Optimization (BCO) is a heuristic optimization method inspired by the foraging behavior of honeybee colonies. Bees use a natural selection process to find food sources and bring them back to the hive.

The BCO uses this principle to solve optimization problems, simulating the behavior of bees to find optimal solutions. Possible solutions are represented by positions in a search space, where each position represents a potential solution [127], [128].

Artificial bees are generated to explore the search space using local and global movements. Local movements are performed around the bee's current position, while global movements are performed by exploring new regions of the search space.

The main steps of the virtual bee algorithm for function optimization are as follows [129] :

- ➢ Creation of a population of virtual bees, each with a solution vector and several parameters to optimize.

- ➢ Coding of optimization functions and conversion to virtual food.

- ➢ Definition of a criterion for transmitting direction and distance to other bees in a manner similar to the physical aptitude of bees (the wriggling dance).

- ➢ Update a population of individuals in new positions for virtual foraging, doing the virtual dance to define distance and direction.

- ➢ Decode the results to obtain the solution to the problem.

These different energy optimization methods offer advantages and disadvantages depending on the complexity and nature of the system to be optimized. Researchers and engineers often choose the most appropriate optimization method according to their objectives and the constraints of their system.

4. Research on energy optimization methods

Researchers S. Bera et al (2014) [130] have proposed, in this paper, an energy-efficient smart metering system to minimize the energy consumption of smart meters for green smart communication network.

In a smart grid, smart meters should be the key technology to support the two-way exchange of information between service providers and end-users. The growing demand for smart meters (residential customers and plug-in hybrid electric vehicles) would result in huge energy consumption, while communicating with smart grid entities. Consequently, green wireless communication technologies should help reduce their impact on the environment.

The researchers incorporated the use of the coalition game formed by multiple coalitions between smart meters to communicate with the service provider. And they showed that there is a stable condition of coalitions for which the gain values of smart meters are maximized. Simulation results show that by using the proposed approach, the energy consumption of smart meters can be reduced, which in turn would enable green wireless communication in the smart grid.

Researchers Z. Abdmouleh et al. (2017) [131] provided an analysis on the environmental, economic, technological, technical and regulatory drivers that have led to the growing interest in distributed generation integration in combination with an overview of the challenges to be overcome. Finally, they reviewed all significant methods applying optimization techniques to the integration of distributed generation from RNG. A summary of common heuristic optimization algorithms with Pro-Con lists is discussed in order to identify potential new avenues for hybrid methods that have not yet been explored.

Researchers Y. Zheng et al. (2018) [132] developed a load transfer algorithm based on economic linear programming with a predictive control model to minimize the operating cost of a microgrid system based on heat and electricity combined with biomass. The model simultaneously manages electrical and thermal energy supply and demand as decision variables. They have also developed an algorithm to optimize load shifting according to renewable energy production and hourly tariff. As an illustrative example, a case study was examined for a conceptual microgrid application connected to the distribution grid in Davis, California. For the assumptions used, the proposed load transfer algorithm improved microgrid performance by modifying the load model and reduced the operating cost by 6.06% and increased the renewable energy fraction by 6.34% compared to the conventional vacuum transfer case. They used Monte Carlo simulation to assess uncertainties between RE, demand and economic assumptions, generating a probability density function for the cost of energy.

Researchers X. Jiang et al (2020) [133] proposed a power scheduling algorithm based on smart home profitability, to improve consumer efficiency and consumption satisfaction. In this proposed method, a definition of profitability for residential power scheduling is introduced. Consumer consumption costs are modeled as a function of electricity payments and user discomfort. The researchers also designed a pair of parameters for a trade-off between user discomfort and their electricity payment, to model profitability as a function of consumer preference. They developed a profitability-based power scheduling algorithm using a fractional programming approach. They analyzed and discussed four consumption models. The results show that this proposed method can effectively improve consumers' consumption efficiency and satisfaction while reducing costs. It is shown that discomfort and extra payment can impact consumers' consumption behavior and smooth the curves of their consumption profile.

Researchers M.M. Rashid et al (2020) [134] have proposed a strategy to improve energy and system cost for residential consumers. They developed an efficient power management algorithm (PMA) with renewable energy systems (RES) and energy storage systems (ESS). They executed the PMA using C++ software. The benefits of RES and ESS with the proposed PMA techniques are analyzed using three scenarios. The researchers carried out extensive case studies to validate the proposed energy efficiency and cost minimization. They demonstrated that the proposed method can save energy and costs by up to 34% and 45% compared with the existing method.

Researchers X. Huang, et al (2021) [135] proposed a wireless sensor network node localization algorithm (MA*-3DDV-Hop) that integrates the improved A* algorithm and the 3DDV-Hop algorithm. In MA*-3DDV-Hop, they first optimized the value of the number of node hops, then corrected the average distance error per hop. Next, they adopted the genetic algorithm of non-dominated sorting with multi-objective optimization (NSGA-II) to locally optimize the coordinates. After selection, crossover and mutation, the Pareto optimal solution is obtained, thus overcoming the problems of premature convergence and poor convergence of existing algorithms. It also reduces coordinate calculation error and increases the localization accuracy of wireless sensor network nodes. For three different multi-peak random scenes, the researchers showed with simulation that the MA*-3DDV-Hop algorithm has better robustness and higher localization accuracy than 3DDV-Hop, PSO-3DDV-Hop, GA-3DDV-Hop and N2-3DDV- Hop.

5. Conclusion

Optimizing energy management within smart power grids represents a central challenge in meeting tomorrow's energy and environmental challenges. Thanks to remarkable advances in the fields of sensors, communications and automation, these intelligent infrastructures offer considerable potential for dynamic, optimal control of energy flows.

As we explored earlier, several optimization methods can be implemented, ranging from scheduling algorithms to machine learning techniques, via multi-objective optimization approaches. Each method has its own advantages and limitations, and the choice of the most appropriate solutions will largely depend on the specific context of the network under consideration.

General conclusion

Excessive use of fossil fuels has led to numerous energy and environmental problems, calling into question the adoption of renewable energies. In order to effectively integrate renewable energies into power grids, a transition to smart grids is essential. In addition to their modeling, which is an essential step due to their complexity, SGs cannot be fully effective without an energy optimization strategy.

Fortunately, significant advances in modeling, simulation and optimization are opening up promising new perspectives. As we have seen, systems approaches, artificial intelligence techniques and multi-objective optimization methods are now making it possible to better understand the complexity of these new-generation power grids, and to considerably improve their management.

More than just technical advances, we're witnessing a veritable revolution in society. The consumer, once a passive user, is now a dynamic player, capable of producing and managing his or her own energy. The traditional lines between production, distribution and consumption are disappearing, paving the way for new economic models and new forms of governance.

Ultimately, it's a vision of a sustainable, resilient and inclusive energy system that is taking shape. A system where cutting-edge technologies combine with new forms of organization to meet the challenges of ecological transition. A system that puts people at the heart of its concerns, in the service of a society that is fairer and more respectful of its environment.

Summary

Dwindling fossil fuel resources, rising global energy demand and the need to reduce the environmental impact of greenhouse gas emissions are major challenges that require effective solutions. Against this backdrop, the development of renewable energies and demand management are emerging as promising alternatives. To meet these challenges, intelligent strategies are needed to manage energy production, regulate its integration into the power grid and maintain a constant balance between supply and demand.

In this book, we began by examining the energy challenges and problems posed by the massive use of fossil fuels. We have highlighted the importance of renewable energies, which represent a radical solution to climate problems and energy shortages. We then explained the need to use hybrid energy systems to overcome the intermittency of renewable energies and guarantee a stable, efficient power supply. A detailed study of these hybrid systems was then carried out.

This hybridization has led to the transformation of conventional power grids into smart grids. To this end, another study of this type of network has been carried out. Given the complexity of these smart grids, it is essential to model them in order to predict and install them correctly. So, in the third chapter, we looked at modeling methods to provide a solid foundation for these techniques.

Talking about smart grids also implies discussing optimal energy management, which is the key to making these grids more efficient and profitable. Therefore, at the end of the book, we have drawn up a bibliographical review of the various energy optimization techniques and algorithms.

This book is just a glimpse into the vast field of future power networks.

Abstract

The decline in fossil energy resources, the increase in global energy demand and the need to reduce the environmental impact of greenhouse gas emissions are major challenges that require effective solutions. In this context, the development of renewable energies and the control of demand appear to be promising alternatives. In order to meet these challenges, intelligent strategies are essential to manage energy production, regulate its integration into the electricity network and maintain a constant balance between supply and demand.

In this work, we began by studying the energy issues and the problems posed by the massive use of fossil fuels. We have thus highlighted the importance of renewable energies, which represent a radical solution to avoid climate problems and lack of energy. We then explained the need to use hybrid energy systems in order to compensate for the intermittency of renewable energies and to guarantee a stable and efficient power supply. A detailed study of these hybrid systems was therefore carried out.

This hybridization has led to a transformation of traditional electricity networks towards intelligent networks. For this, another study on this type of network was carried out. Given the complexity of these smart power grids, modeling them is essential to predict and install them correctly. So, in the third chapter, we studied modeling methods to provide a solid foundation on these techniques.

Talking about smart grids also involves discussing optimal energy management, which is the key to making these grids more efficient and profitable. Consequently, at the end of the work, we have developed a bibliographic study on the different techniques and algorithms for energy optimization.

This work is only an overview of a very broad field relating to the electrical networks of the future.

References

References

[1] Presentation of renewable energies, available at: http://www.energies-renouvelables.org, [accessed January 2022].

[2] J. El Khaldi, L. Bouslimi, H. Wertani & M.N. Lakhoua, "Overview on modeling and management of Smart Grids", Independent Journal of Management & Production (IJM&P), v. 12, n. 5, July-August 2021.

[3] P. Asantewaa Owusu & S. Asumadu-Sarkodie, "A review of renewable energy sources, sustainability issues and climate change mitigation", Cogent Engineering, 3:1, 2016.

[4] La production d'électricité d'origine renouvelable dans le monde, Observatoire des énergies renouvelables, Rapport technique, available at: http://www.energies-renouvelables.org/observ-er/html/inventaire/Fr/preface.asp, [consulted February 2021].

[5] M. Mouhamed AL Anfaf, "Contribution à la modélisation et l'optimisation de systèmes énergétiques multi-sources et multi-charges", PhD thesis, Ecole Doctorale "Sciences et Ingénierie Ressources Procédés Produit Environnement" de l'Université de Lorraine, July 2016.

[6] Y. Saleem, N. Crespi, M. Husain Rehmani & R. Copeland, "Internet of things-aided Smart Grid: technologies, architectures, applications, prototypes and future research directions", IEEE, 2019.

[7] Y. Yan, Y. Qian, H. Sharif & D. Tipper, " A Survey on Smart Grid Communication Infrastructures: Motivations, Requirements and Challenges" (2013). Faculty Publications from the Department of Electrical and Computer Engineering. 316.5

[8] M. Remy Rigo-Mariani, "Méthodes de conception intégrée " dimensionnement-gestion " par optimisation d'un micro-réseau avec stockage", PhD Thesis, Institut National Polytechnique de Toulouse, Plasma and Energy Conversion Laboratory, December 2014.

[9] IEA, "Global Energy & CO_2 , Status Report 2019", Available online: https://www.iea.org/reports/global-energy-co2-status-report-2019, [accessed March 2022].

[10] Aysar M.M. Yasin, "Distributed Generation Systems Based on Hybrid Wind/Photovoltaic/Fuel Cell structures", PhD Thesis, Università Degli Studi di Catania Scuola Superiore di Catania 2012.

[11] IRENA, "Renewable Energy Statistics 2018", Available online: https://irena.org/-/media/Files/IRENA/Agency/Publication/2019/Jul/IRENA_Renewable_energy _statistics_2018.pdf, [consulted in january 2021].

[12] IRENA, "Renewable Energy Statistics 2019". Available online: https://irena.org/-/media/Files/IRENA/Agency/Publication/2019/Jul/IRENA_Renewable_energy _statistics_2019.pdf, [consulted in april 2022].

[13] M. Huynh Quang, "Optimisation de la production de l'électricité renouvelable pour site isolé", PhD thesis, Université de Reims Champagne-Ardenne, Ecole doctorale STS, 2013.

[14] E. Observ, "La production d'électricité d'origine renouvelable dans le monde", Collection chiffres et statistiques. Ninth Inventory-Edition, 2011.

[15] X. Wen, "Stochastic optimization for generation scheduling in a local energy community under renewable energy uncertainty", Thesis, Centrale Lille, Ecole doctorale SPI072, December 2020.

[16] M. Dahmane, "Gestion, optimisation et conversion des énergies pour habitat autonome", These de Doctorat, Université de Picardie Jules Verne, 2015.

[17] L. Stoyanov, "Etude de Différentes Structures de Systèmes Hybrides à Sources D'énergie Renouvelables", PhD Thesis, Sofia Technical University ,2011.

[18] C. Ashari, M. Nayar & C.V. Keerthipala, "Optimum Operation Strategy and Economic Analysis of a Photovoltaic - Diesel - Battery -Mains Hybrid Uninterruptible Power Supply", Renewable Energy, Vol.22, Issues 1-3, pp.247- 254, March 2001.

[19] T. Zhou, "Commande et supervision énergétique d'un générateur hybride actif éolien incluant du stockage sous forme d'hydrogène et des super-condensateurs pour l'intégration dans le système électrique d'un micro réseau", PhD thesis, Ecole Centrale de Lille 2009.

[20] D. Abbes, "Contribution au dimensionnement et à l'optimisation des systèmes hybrides éoliens-photovoltaïques avec batteries pour l'habitat résidentiel autonome", PhD thesis, Ecole doctorale des sciences et ingénierie pour l'information, June 2012.

[21] M. Dugay, "Wind and solar energy resources in Quebec", IREC research report, Institut de recherche en économie contemporaine, December 2010.

[22] B. Abdelhalim, "Etude et optimisation d'un multi système hybride de conversion d'énergie éléctrique", Université Constantine1, Faculté des sciences de la technologie, March 2015.

[23] L. Miguel, "Contribution à l'optimisation d'un système de conversion éolien pour une un unite de production isolée," École Doctorale des Sciences et Technologies de l'Information des Télécommunications et des Systèmes, Université Paris Sud, 2008.

[24] L. Croci, "Gestion de l' énergie dans un systeme multi-sources photovoltaique et éolien avec stockage hybride batteries/supercondensateurs", PhD Thesis, University of Poitiers, 2014.

[25] P. Bajpai & V. Dash, "Hybrid renewable energy systems for power generation in stand-alone applications: A review", Renewable and Sustainable Energy Reviews, vol. 16, no.5, pp.2926-2939, 2012.

[26] M. Ashari & C.V. Nayar, "An optimum dispatch strategy using set points for a photovoltaic (pv)-diesel-battery hybrid power system", Solar Energy, 66(1) :1 - 9, 1999.

[27] V. Sark, D. Cocard, P. Beutin, G. Merlo, B. Mohanty, J. van den Akker, A. Idris, A. Firag & A. Waheed, "The first PV-diesel hybrid system in the Maldives installed at Mandhoo Island", 21st European 178 Photovoltaic Solar Energy Conference, 4-8 September 2006, Dresden, Germany, pp. 3039-3043, 2006.

[28] A. Adiyabat & K. Kurokawo, "Photovoltaic systems for village electrification in Mongolia: Techno-economics analysis of hybrid systems in rural community centers", The International PVSEC-14, Bengkok, Tailand, 2004.

[29] R. Abhinav & N.M. Pindoriya, "Grid Integration of Wind Turbine and Battery Energy Storage System: Review and Key Challenges", IEEE: Power Systems, 2016.

[30] A. Mahesh & K. Singh Sandhu, "Hybrid wind/photovoltaic energy system developments: Critical review and findings", Renewable and Sustainable Energy Reviews 52 (2015) 1135-1147.

[31] D. Saheb-Koussa, M. Haddadi & M. Belhamel, "Economic and technical study of a hybrid system (windephotovoltaicediesel) for rural electrification in Algeria", Appl. Energy 86 (2009) 1024e1030.

[32] D. Saheb-Koussa, M. Koussa, M. Haddadi & M. Belhamel, "Hybrid options analysis for power systems for rural electrification in Algeria", Energy Procedia 6 (2011) 750e758.

[33] A. Haghighat Mamaghani, S. Alberto Avella Escandon, B. Najafi, A. Shirazi & F. Rinaldi, "Techno-economic feasibility of photovoltaic, wind, diesel and hybrid electrification systems for off-grid rural electrification in Colombia", Renewable and Sustainable Energy Reviews, Elsevier, 2016.

[34] M. Junaid Khan, A. Kumar Yadiw & L. Mathew, "Techno economic feasibility analysis of different combinations of PV-Wind-Diesel-Battery hybrid system for telecommunication applications in different cities of Punjab, India", Renewable and Sustainable Energy Reviews, Volume 76, September 2017, Pages 577-607.

[35] S. Malu, "Modelling, Control and Simulation of a Microgrid based on PV System, Battery System and VSC", Thesis, Escola Tècnica Superior d'Enginyeria Industrial de Barcelona, 2018.

[36] R.H. Lasseter, "Micro-grids" (distributed power generation), IEEE Power Engineering Society Winter Meeting, Vol.01, pp.146-149, Columbus, Ohio, Feb 2014.

[37] R.H. Lasseter, "Micro-grids", IEEE Power Engineering Society Winter Meeting, Vol.01, pp. 305- 308, New York, NY, 2015.

[38] S. Ali, Z. Zheng, M. Aillerie, J.P. Sawicki, M.C. Péra & D. Hissel, "A Review of DC Microgrid Energy Management Systems Dedicated to Residential Applications". Energies 2021, 14, 4308.

[39] E. Planas, J. Andreu, J.I. Gárate, I. Martínez de Alegría & E. Ibarra, "AC and DC Technology in Microgrids: A Review". Renew. Sustain. Energy Rev. 2015, 43, 726-749.

[40] A. Egea, A. Junyent, O. Gomis, "Active and Reactive Power Control of Grid Connected Distributed Generation Systems", Green Energy and Technology. Heidelberg : 2011, pp 47-81.

[41] General microgrids, what is a microgrid, Availble online: https://www.generalmicrogrids.com/about-microgrids, [accessed November 2021].

[42] A. Houari, "Contribution à l'étude de micro-réseaux autonomes alimentés par des sources photovoltaïques", PhD thesis, University of Lorraine, 2012.

[43] H. Fakham & B. Francois, "Power Control Design of a Battery Charger in a Hybrid Active PV Generator for Load-Following Applications", IEEE Transactions on Industrial Electronics, vol. 58, no. 1, pp. 85 - 94, Jan. 2011.

[44] G-Y. Choe, J-S. Kim, B-K. Lee, C-Y. Won & T-W. Lee, "A Bi-directional Battery Charger for Electric Vehicles Using Photovoltaic PCS Systems", in: Vehicle Power and Propulsion Conference (VPPC), 2010, pp. 1 - 6, Sept. 2010.

[45] M. Vandenbergh, S. Beverungen, B. Buchholz, H. Colin, N. Ketjoy, F. Kininger, D. Mayer, J. Merten, J. Reekers, P. Strauss1, T. Suwannakum & X.Vallvé, "Expandable Hybrid Systems for Multi-User MiniGrids", in: Use of Electronic-Based Power Conversion for Distributed and Renewable Energy Sources, 2008.

[46] X. Wu, Z. Wang, T. Ding & Z. Li, "Hybrid AC/DC Microgrid Planning with Optimal Placement of DC Feeders". Energies 2019, 12, 1751.

[47] N. Yang, B. Nahid-Mobarakeh, F. Gao, D. Paire, A. Miraoui & W. Liu, "Modeling and Stability Analysis of Multi-Time Scale DC Microgrid", Electr. Power Syst. Res. 2016, 140, 906-916.

[48] H. Wertani, "Contribution à l'analyse et à la modélisation systémique d'un réseau électrique intelligent", PhD thesis, Université de Carthage, Ecole Nationale d'Ingénieurs de Carthage, Laboratoire de recherche électricité intelligente & TIC, 2022.

[49] M. Chebbo, "EU smart grids framework: electricity networks of the future 2020 and beyond", in Proc. IEEE Power Engineering Society General Meeting, Tampa, FL, June 2007, pp.1 - 8

[50] S. Arup, S. Neogi, R.N. Lahiri, S. Chowdhury, S.P. Chowdhury & N. Chakraborty, "Smart Grid initiative for power distribution utility in India", IEEE Power and Energy Society General Meeting, 2011.

[51] E. Santacana, B. Husain, F. Pinnekamp & P. Halvarsson, "Let there be intelligent light - the future highways of clean, safe and sustainable electricity", Power Grids of the Future, Vol.1, N°10, 2010.

[52] I. Tegani, "Optimisation et contrôle d'un micro Smart Grid utilisant une pile à combustible, des supercondensateurs, des batteries, une éolienne et une source photovoltaïque", PhD Thesis, Université Mohamed Khider Biskra, Faculté des sciences et de la technologie, 2016.

[53] B. Spandana, "Smart Grid Technologies for Efficiency Improvement of Integrated Industrial Electric System", *University of New Orleans Theses and Dissertations*. 115, 2011.

[54] R. Abdelouahab & B. Smail, Study of SMES energy storage in smart grids, 2016.

[55] R. Glaa, "Sur la modélisation, le contrôle et la supervision d'un réseau intelligent d'électricité", PhD thesis, Université de Carthage, Ecole Nationale d'Ingénieurs de Carthage, Laboratoire de recherche électricité intelligente & TIC, 2019.

[56] R. M. Badreddine, "Gestion énergétique optimisée pour un bâtiment intelligent multi sources multi-charges : différents principes de validations", PhD thesis, University of Grenoble, 2012.

[57] G. Guillaume, "Optimisation de la diffusion de l'energie dans les smart grid", PhD thesis, Université de Versailles-Saint Quentin en Yvelines, 2014.

[58] O. Mosbahi & M.Khalgui, "New solutions for optimal power productions, distribution and consumption in smart grids", International Journal Modelling, Identification and Control, Vol.26, no. 2, 2016.

[59] Y. Kabalu, "A survey on smart metering and smart grid communication", Renewable and

sustainable Energy Reviews, 2016.

[60] S. Arup, S. Neogi, R.N. Lahiri, S. Chowdhury, S.P. Chowdhury & N. Chakraborty, "Smart

Grid initiative for power distribution utility in India", IEEE Emerging trends in Electrical and

Electronics Engineering, 2011.

[61] S. Arup, S. Neogi, R.N. Lahiri, S. Chowdhury, S.P. Chowdhury & N. Chakraborty, "Smart Grid initiative for power distribution utility in India", IEEE Power and Energy Society General Meeting, Detroit, MI, USA, pp. 1-8, doi: 10.1109/PES.2011.

[62] S.R. Depuru et al, "Smart metres for power grid, Challenges issues advantages status", IEEE PES Power Systems Conference and Exposition, 2011.

[63] M.L. Tuballa & M.L. Abundo, "A review of the smart grid technologies", Renewable and Sustainable Energy Reviews, 2016.

[64] P. Siano, "Demand response and smart grid-A survey", Renewable and Sustainable Energy Reviews, pp.461-478, 2014.

[65] C. Neureiter, M. Uslar, D. Engel & G. Lastro, "A standards-based approach for domain specific modelling of smart grid system architectures, "11th System of Systems Engineering Conference, 2016.

[66] H.A. M. Abumeteir, "A proposed SCADA system to improve the conditions of the electricity sector in Gaza Strip", Thesis,The Islamic University Of Gaza, 2012.

[67] M. Ourahou, W. Ayrir & B. EL Hassouni, "Review on smart grid control and reliability in presence of renewable energies: Challenges and prospects," Mathematics and Computers in Simulation, 2018.

[68] L. Hong, Z. Hong-lei, C. Wen-ting, Y. Jian-cheng, Z. Shi-yu & W. Bo, "Analysis of structure of intelligent park multi-level energy transfer model," TENCON 2015 - 2015 IEEE Region 10 Conference, Macau, 2015, pp. 1-6,

[69] A. Chabaud, "Micro-réseau intelligent pour la gestion des ressources énergétiques", PhD thesis, Ecole doctorale énergie environnement Perpignan, 2014.

[70] Y. Cunjiang, Z. Huaxun & Z. Lei, "Architecture design for smart grid", Energy Procedia 2012, 17, 1524-1528.

[71] S. Goel & D. Bakken, "Smart Grid of the future: vision for year 2030", IEEE Vision for Smart Grid Control: 2030 and Beyond Roadmap. IEEE; Piscataway, NJ, USA: 2013. pp. 1-12.

[72] M.E. El Hawary, "The smart Grid: State of the art and future trends", Electric Power Components and Systems, 42(3-4):239-250, 2014.

[73] M.L. Tuballa & M.L. Abundo, "A review of the development of Smart Grid technologies", Renewable and Sustainable Energy Reviews, Elsevier, Volume 59, June 2016, Pages 710-725.

[74] S. Howell, Y. Rezgui, J.L. Hippolyte, B. Jayan & H. Li, "Towards the next generation of smart grids: Semantic and holonic multi-agent management of distributed energy resources", Renew. Sustain. Energy Rev. 2017, 77, 193-214.

[75] S. Kakran & S. Chanana, "Smart operations of smart grids integrated with distributed generation: A review", *Renew. Sustain. Energy Rev.* 2018, *81*, 524-535.

[76] Y. Ming, X. Zhang & X. Shen, "Efficient Privacy-Preserving Multi-Dimensional Data Aggregation Scheme in Smart Grid," IEEE Access. 2019; 7:32907-32921.

[77] D. Tang, Y.P. Fang, E. Zio & J.E. Ramirez-Marquez, "Resilience of Smart Power Grids to False Pricing Attacks in the Social Network" IEEE Access. 2019; 7:80491-80505.

[78] K. Nasraoui, M.N. Lakhoua, & L. El Amraoui, L, "Study and analysis of micro smart grid using the modeling language SysML". In 2017 International Conference on Green Energy Conversion Systems (GECS) (pp. 1-8). IEEE, March 2017.

[79] P. Bernus, O. Noran & J. Riedlinger, "Using the Globemen Reference Model for Virtual Enterprise Design in After Sales Service", CiteSeerx: Scientific Literature Digital Library and Search Engine, 15, 2007.

[80] J. El Khaldi, L. Bouslimi, A. Balti & M.N. Lakhoua, "Study and analysis of the methods of the enterprise modeling", Design Engineering, issue: 1/pages: 2945/2948/ 2022.

[81] M. Rahmouni & M.N.Lakhoua, "using function and decision models for enterprise restructuring", STA, monastir, 2010.

[82] H. Mathieu. "Modélisation conjointe de l'infrastructure et des processus pour l'administration pro-active de l'entreprise distribuée", Doctoral thesis, Ecole Doctorale informatique et information pour la société, 2004.

[83] S. Galland, "Approche multi-agents pour la conception et la construction d'un environnement de simulation en vue de l'évaluation des performances des ateliers multi-sites", PhD thesis, Saint-Etienne, 2001.

[84] J. Ryan & C. Heavey, "Process modeling for simulation", Elsevier, computers in industry, 14, 2006.

[85] J. Shin, J. Joo, J. Choi, S.H. Han & C. Hyunbo, "A prototype virtual reality system through IDEF modeling for product configuration and analysis", International Journal of Industrial engineering, 11, 2000.

[86] M. Lauras, "Méthodes de diagnostic et d'évaluation de performance pour la gestion de chaine logistique", Toulouse: Institut National Polytechnique, 2004.

[87] S. Sperandio, "Usage de la modélisation multi-vue d'entreprise pour la conduite des systèmes de production", PhD thesis, Bordeaux 1, 2009.

[88] V. Augusto, "Modélisation, analyse et pilotage de flux en milieu hospitalier à l'aide d'UML et des réseaux de Petri", PhD thesis, Saint-Etienne, 2008.

[89] A. Demri, A. Charki, F. Guérin, P. Kahn & H. Christofol, "Analyse qualitative et quantitative d'un système mécatronique", 4th International Conference on Computer Integrated Manufacturing, (p. 14), 2007.

[90] R. Abbou, Z. Simeu-Abazi, & M. Di-Mascolo. "Petri nets for modeling and analyzing the performance of a maintenance workshop", MOSIM, (p. 6). Toulouse, 2003.

[91] J. Ben Salem, "Contribution à la modélisation systémique des systèmes mécatroniques. Cas d'un système de freinage ABS", PhD thesis, Carthage University, Carthage National Engineering School, 2018.

[92] M. Elhamdi, "modélisation et simulation des chaines de valeurs en entreprise - une approche dynamique des systèmes et aide à la décision", PhD thesis, Paris: Ecole Centrale, 2005.

[93] A. Hassan, "Proposition et développement d'une approche pour la maîtrise conjointe qualité/coût lors de la conception et de l'industrialisation du produit", PhD thesis, Metz: École Nationale Supérieure d'Arts et Métiers, 2010.

[94] H. Kromm & J. Christophe Deschamps, "Modélisation de processus pour une évaluation par niveaux de détail successifs", Conférence francophone de modélisation et de simulation. Troyes (France), 2006.

[95] P ; Boutin, "Définition d'une méthodologie de mise en œuvre et de prototypage d'un progiciel de gestion d'entreprise (ERP). Modélisation et simulation", PhD thesis, Ecole Nationale Supérieure des Mines de Saint-Etienne, 2001.

[96] F. Drras, "Proposition d'un cadre de référence pour la conception et l'exploitation d'un progiciel de gestion intégré", PhD thesis, Toulouse: Institut national polytechnique de toulouse, 2004.

[97] M. Bennour, "Contribution à la Modélisation et à l'Affectation des Ressources Humaines dans les Processus", Montpellier II, 2004.

[98] A.P. Sage & W.B. Rouse. "Handbook of system engineering and management", USA: Library of Congress Cataloging-in-publication data, (2009).

[99] B. henderson, A.L. Parc.Lacayrelle & J.M.VBruel, "formalization of the whole - part relationship in the unified Modeling lnguage", IEEE transactions on software Engineering, 2003.

[100] M. Aouag, "Des diagrammes UML 2.0 vers les diagrammes orientés aspect à l'aide de transformation de graphes", PhD Thesis, University of Constantine, Algeria, 2014.

[101] K.D. Nguyen, Z. Sun & P.S. Thiagarajan, "Model-driven design via executable UML to System", School of computing, national university of Singapore, Real-Time Systems Symposium5-8 Dec, pp. 459 - 468, 2004.

[102] S. Turki, "Application du standard IEEE 15288, de l'architecture MDA et du langage SysML à la conception des systèmes mécatroniques", PhD thesis, Marseille III, 2008.

[103] B. Vallespir, V. Chapurlat, & C. Brae, "L'intégration en modélisation d'entreprise : les chemins d'U.E.M. L ", MOSIM, (p. 6). Toulouse, 2003.

[104] K. Mertins, & R. Jochem, "Architectures, methods and tools for enterprise engineering", Elsevier, International journal of production economics, 10. 2005.

[105] A. Talbi, A. Hammouche, & C. Tahon, "Analyse de l'entreprise dans une démarche d'intégration", (p. 33). France: APII - JESA, 2002.

[106] J. TOUZI, "Conception d'un système d'information collaboratif dans une approche orienté services", congré, (p. 6). Toulouse, 2006.

[107] A. Abdmouleh, "Composants pour la modélisation des processus métier en productique basés sur CIMOS", PhD thesis, Ile du Saulcy: LGIPM - ENIM, 2004.

[108] A. Zaidat, "Spécification d'un cadre d'ingénierie pour les réseaux d'organisations", PhD thesis, Saint-Etienne: Ecole Nationale Supérieure des Mines, 2005.

[109] A.K. Traoré, "Gestion d'un système autonome hybride photovoltaïque éolien pour application agricoles", PhD thesis, Université du Québec à Trois-Rivières, July 2016.

[110] R. Chedid & S. Rahman, "Unit sizing and control of hybrid", IEEE Transactions on Energy Conversion, vol. 12, no. %11, pp. 79-85, 1997.

[111] Y. Gaoua, "Mathematical models and nonlinear and combinatorial optimization techniques for multi-source system energy management: towards a real-time implementation for different hybrid vehicle electrical structures", PhD thesis, Institut National Polytechnique de Toulouse (INP Toulouse), 2014.

[112] I. Vechiu, "Modélisation et analyse de l'intégration des énergies renouvelables dans un réseau autonome", PhD thesis, University of Le Havre, 2010.

[113] O.H. Mohamed, Y. Amirat, M. Benbouzid, "Particle swarm optimization of a hybrid wind/tidal/PV/battery energy system. Application to a remote area in Bretagne, France", ScienceDirect, Energy Procedia, 162: 87-96, 2019.

[114] A. Mills, S. Al-Hallaj, "Simulation of hydrogen-based hybrid systems using Hybrid2", International Journal of Hydrogen Energy 29:991-999, 2004.

[115] N.M. Kumar, M.R. Kumar, P.R. Rejoice & M. Mathew, "Performance analysis of 100 kWp grid connected Si-poly photovoltaic system using PVsyst simulation tool", ScienceDirect: Energy Procedia 117: 180-189, 2017.

[116] J. Mast, S. Raidle, J. Gerlach & O. Bringmann, "Development of a modelling and simulation methodology for hierarchical energy system scenarios," International Journal of Smart Grid and Clean energy, 2019.

[117] M. Cavazzuti, "Optimization Methods: From Theory to Design Scientific and Technological Aspects in Mechanics", Springer Science & Business Media, 2012.

[118] A.F. Izmailov & M.V. Solodov, "Newton-type methods for optimization problems without constraint qualifications", Society for Industrial and Applied Mathematics: Vol. 15, No. 1, pp. 210-228, 2004.

[119] E. Polykarpou & E. Kyriakides, "Parameter estimation for measurement-based load modeling using the levenberg-marquardt algorithm", 18th Mediterranean Electrotechnical Conference (MELECON), Lemesos, 2016.

[120] G.A. Mary & R. Rajarajewari, "Smart Grid cost optimization using genetic algorithm", IJRET: International Journal of Research in Engineering and Technology, Volume: 03 Special Issue: 07, May-2014.

[121] E.Lutton, "Darwinisme artificiel", INRIA Rocquencourt Equipe Complex Proj et Fractales, 2004.

[122] K. Ouramdane, "Modélisation et optimisation du système d'excitation du groupe turbine-alternateur du simulateur analogique d'Hydro-Québec", Université du Québec en Abitibi-Témiscamingue, May 2015.

[123] J.M. Alliot & N. Durand, "Algorithmes génétiques," Centre d'Etudes de la Navigation Aérienne, 2005.

[124] H. Rathore, "Mapping Biological Systems to Network Systems", ed: Springer, 2016.

[125] M. Dorigo and L. M. Gambardella, "Ant colony system: a cooperative learning approach to the traveling salesman problem", Evolutionary Computation, IEEE Transactions on, vol. 1, pp. 53-66, 1997.

[126] Z. Chi Su Su Hlang & M. Aye Khin, "Solving traveling salesman problem by using improved ant colony optimization algorithm", International Journal of Information and Education Technology, Vol. 1, No. 5, December 2011.

[127] D. Teodorovié, P. Lucié, G. Markovi & M. Dell orca, "Bee colony optimization: principles and applications", 8th Seminar on Neural Network Applications in Electrical Engineering, NEUREL-2006.

[128] D. Teodorovic, M. Dell' Orco, "Bee Colony Optimization-A cooperative learning approach to complex transportation problems" in Abstracts - of 10th EWGT Meeting and 16th Mini EURO Conference, Poznan, 2005.

[129] S. Mouassa, "Optimisation de l'écoulement de puissance par une méthode métaheuristique (technique des abeilles) en présence d'une source renouvelable (éolienne) et des dispositifs FACTS", PhD thesis, Université Ferhat Abbas de Sétif 1, 2012.

[130] S. Bera, S. Misra & M.S. Obaidat, "Energy-efficient smart metering for green smart grid communication", Proceedings of the 2014 IEEE Global Communications Conference; Austin, TX, USA. 8-12 December 2014; pp. 2466-2471.

[131] Z. Abdmouleh, A. Gastli, L. Ben-Brahim, M. Haouari & N.A. Al-Emadi, "Review of optimization techniques applied for the integration of distributed generation from renewable energy sources", Renew. Energy 2017, 113, 266-280.

[132] Y. Zheng, B.M. Jenkins, K. Kornbluth, A. Kendall & C. Traeholt, "Optimization of a biomass-integrated renewable energy microgrid with demand side management under uncertainty", Applied Energy, Elsevier, Volume 230, 15 November 2018, Pages 836-844.

[133] X. Jiang & L. Wu, "Residential Power Scheduling Based on Cost Efficiency for Demand Response in Smart Grid," IEEE Access. 2020; 8:197324-197336.

[134] M.M. Rashid , M.A. Hossain, R. Shah, M.S. Alam, A.K. Karmaker & M. Rahman. "An Improved Energy and Cost Minimization Scheme for Home Energy Management (HEM) in the Smart Grid Framework," Proceedings of the 2020 IEEE International Conference on Applied Superconductivity and Electromagnetic Devices (ASEMD), Tianjin, China. 16-18 October 2020; pp. 1-2.

[135] X. Huang , D. Han, M. Cui, G. Lin &X. Yin, "Three-Dimensional Localization Algorithm Based on Improved A* and DV-Hop Algorithms in Wireless Sensor Network", *Sensors.* 2021; 21:448.

I want morebooks!

Buy your books fast and straightforward online - at one of world's fastest growing online book stores! Environmentally sound due to Print-on-Demand technologies.

Buy your books online at
www.morebooks.shop

Kaufen Sie Ihre Bücher schnell und unkompliziert online – auf einer der am schnellsten wachsenden Buchhandelsplattformen weltweit! Dank Print-On-Demand umwelt- und ressourcenschonend produziert.

Bücher schneller online kaufen
www.morebooks.shop

Printed by Books on Demand GmbH, Norderstedt / Germany